HARD CORPS

LEGENDS OF THE MARINE CORPS

By MSgt A. A. Bufalo USMC (Ret)

Copyright © 2004 by S&B Publishing

ISBN 978-0-9745793-5-1

First Printing – 10 November 2004
Printed in the United States of America

www.AllAmericanBooks.com

HARD CORPS ~ Legends of the Corps

OTHER BOOKS BY ANDY BUFALO

SWIFT, SILENT & SURROUNDED
Sea Stories and Politically Incorrect Common Sense

THE OLDER WE GET, THE BETTER WE WERE
MORE Sea Stories and Politically Incorrect Common Sense
Book II

NOT AS LEAN, NOT AS MEAN, STILL A MARINE!
Even MORE Sea Stories and Politically Incorrect Common Sense
Book III

THE ONLY EASY DAY WAS YESTERDAY
Fighting the War on Terrorism

TO ERR IS HUMAN, TO FORGIVE DIVINE
However, Neither is Marine Corps Policy
A Book of Marine Corps Humor

AMBASSADORS IN BLUE
In Every Clime and Place
Marine Security Guards Around the World

FOR

Major G. Warren Dickey USMC

The last of the meat-eaters...

PREFACE

One of the things which makes the Marine Corps different (and better) is we know our history. Ask *any* Marine where the Corps was founded, who the first Commandant was or the year of our birth, and he will tell you "Tun Tavern, Captain Samuel Nicholas and 1775" without hesitation. That is because we use our history as a foundation, upon which the Marines of each succeeding generation build.

Several excellent, comprehensive histories of the Marine Corps have been written, and this book is not meant to compete with them. *Hard Corps* is simply meant to be a tribute to some of the legendary Marines and battles which have built the Corps' reputation as the world's premier fighting force.

Naturally, it was impossible to include *all* of the heroes and battles from our illustrious history. To do so would have required a volume the size of a phone book. Also absent are some of the great leaders and personalities like General John A. Lejeune and bandleader John Philip Sousa. Their stories can be told on another day. What you *will* find are tales of extraordinary heroism, dedication and loyalty.

Our history is important to us. It's part of who we are. So the next time you hear someone ask, "Why do Marines think they are so special?" just hand them a copy of this book and let them read about some of our legends. I guarantee they will pause at the end of each chapter and say, "Damn! That was *Hard Corps*!"

TABLE OF CONTENTS

THE SHORES OF TRIPOLI

Lieutenant Presley N. O'Bannon

"Old breed? New breed? There's not a damn bit of difference so long as it's the *Marine* breed!"
- Lieutenant General Lewis B. Puller

On a lonely knoll in a Frankfurt, Kentucky cemetery stands a simple stone marking the grave of the "Hero of Derne." It is among the final resting places of vice-presidents, senators, governors, artists, and scores of local patriots who fell in action taming the wilderness and fighting foreign aggressors.

The story of Lieutenant Presley Neville O'Bannon begins in 1805. For several years American ships plying the waters along the coast of North Africa had been endangered by bands of Barbary pirates who demanded payments be made to the area's many rulers in exchange for the "protection" of American lives and shipping.

The Barbary Coast states of Morocco, Algiers, Tunis and Tripoli (today known as Libya) were run by high-seas pirates. Though no one had yet coined the phrase, they were certainly the "Axis of Evil" of their day. For many centuries the ships of these four North African countries had been the scourge of shipping in that vital body of water. Often seizing vessels and holding their crews for ransom, or even selling them into slavery, these countries were so intimidating that most European nations paid yearly "tributes" to them in order to secure right of passage for their shipping.

Prior to gaining independence from Great Britain, American ships had not fretted, since they had come under the protection of the powerful British Navy. The sultans and emirs of the Barbary Coast were ready to challenge most navies, but the British Navy was an exception. The end of that British protection was the signal for the pirates to begin exploiting the shipping of the new, relatively weak United States. This they did, and in 1799 the U.S. government gave in and agreed to pay each of the Barbary Coast countries $18,000 a year for the privilege of sailing in "their" sea.

This agreement broke down in 1801, however, as the Pasha of Tripoli, Yusat Karamanli, demanded more money - a demand the U.S. government refused. The Pasha declared war, and an American fleet was sent to the Mediterranean to deal with him. "Millions for defense, but not one cent for tribute" was their battle cry. This slogan conveniently overlooked the fact that we were still paying tribute to the *other* Barbary Coast countries. Then in October of 1803 a ship of the U.S. fleet, the frigate *USS Philadelphia*, ran aground and was captured. The crew ended up in the Pasha's dungeon.

Naval agent General William Eaton proposed that the United States ally itself with the Pasha's disgruntled older brother Hamet and send a force overland from Alexandria, Egypt to attack Tripoli. Hamet commanded a force of about five hundred men, most of whom were Egyptian Mameluke tribesmen. If successful, the force could free the prisoners and also perform a "regime change" in Tripoli, deposing the present Pasha and replacing him with Hamet.

Although the U.S. was tired of a naval war which had dragged on for several years, it was decided to carry out the plan and carry the fight to Derne, the inland stronghold of the enemy and chief fortress of Tripoli. To do this, General Eaton asked for one hundred Marines from a nearby U.S. naval squadron. In answer to his request, a young twenty-nine-year-old Virginian named Lieutenant Presley Neville O'Bannon and seven enlisted Marines from aboard *USS Argus* were placed at Eaton's disposal.

O'Bannon was given an odd assortment of men with which to form a task force formidable enough to seek the surrender of Jussup, the reigning Bey of Tripoli. Finally, on March 8, 1805, his handful of Marines, some Turks, a few Greek mercenaries, and the motley crew of cut-throats and sheiks loyal to Hamet started from Egypt on the 600-mile trek across the desert of Barca.

Along the way, every obstacle known to the Middle East beset Eaton and O'Bannon. Instead of the usual two weeks, the trip covered forty-five days. Many times O'Bannon was called upon to prevent the Muslims from plundering the Christians, and it was he who brought the numerous revolts of the camel drivers to a halt. The Marine lieutenant constantly prodded the Arab chiefs, who repeatedly refused to proceed. All of these delays prolonged the journey, stretched food rations, and at times exhausted water supplies.

At long last on April 25 they arrived at Derna. Surely by then many in this small army must have been happy at the prospect of battle, as opposed to dying a miserable death in the desert. A message was sent to the governor of Derna to surrender. His defiant reply was, "My head or yours." Shortly after this, the attacking force was bolstered by the arrival of *USS Argus*, *USS Hornet*, and *USS Nautilus* in the harbor.

It was decided that Hamet and his Mamelukes would attack the governor's castle, while O'Bannon and his Americans, along with the Greeks and Turks, would lead an assault on the harbor fort. The naval guns would assist by bombarding the objectives.

As the attack began the firing from the governor's castle proved too much for Hamet's force, and they held back. With enemy reinforcements known to be on the way, the attackers were in dire need of a quick victory. Eaton ordered O'Bannon to lead his men in a frontal assault on the harbor fort. Two hours of desperate fighting ensued, and finally O'Bannon and his men made a daring bayonet charge and drove the Tripolitans from the fort and captured the enemy guns there before they could be spiked. This would prove to be important.

O'Bannon had carried a U.S. flag with him and now, for the first time in history, the Stars and Stripes were raised over foreign soil. Seeing this, the defenders in the governor's castle took flight and Hamet's men took possession of it and the town. The victory was not complete however, for now the feared enemy reinforcements arrived - determined to recover what had been lost.

A number of vigorous assaults followed. All were repulsed, with O'Bannon's men able to use the captured guns of the fort to good effect. Finally the Tripolitans gave up, and the battle of Derna was over. Presley O'Bannon had led the first victory of American land forces on foreign soil, but it had not come without a cost. He had lost thirteen killed in the attack, including two of his Marines: Privates John Whitten and Edward Steweard. The people of the town proclaimed Hamet the new ruler of Tripoli, but the victory was fleeting.

Hamet Karamanli promptly took over as ruler of Tripoli and presented the Marine lieutenant with his personal jeweled sword, the same type used by his Mameluke tribesmen. Today's Marine officers still carry this type of sword, with its ivory hilt and gold eagle head, in commemoration of the Corps' service during the Tripolitian War. O'Bannon's exploits would be commemorated by the Marine Corps and the Navy as well... and appropriately, the actions of O'Bannon and his small group of Marines are commemorated in the second line of the Marines' Hymn with the words, "To the Shores of Tripoli." These same words were also inscribed across the top of the Marine Corps' first standard.

Unfortunately for O'Bannon and especially Eaton, who had been the mastermind behind the plan to eliminate the Pasha, the politicians did not manage to hold on to the gains won by the blood of their Marines. Jefferson's government negotiated a treaty which did include the release of all the American sailors imprisoned there, but it still required the United States to pay in order to ensure no further attacks on its shipping. And worst of all, it left the Pasha in power.

Eaton was outraged, and Hamet was forced into exile in Sicily.

Thus the fruits of the victory were tainted, but the skill and courage displayed by O'Bannon and his men were unquestionable. In addition, they had demonstrated that the new American nation would stand up for its rights, with its armed forces, if need be.

Upon his return to this country O'Bannon was given a welcome by the people of Philadelphia and acclaimed as "The Hero of Derne." After his separation from service he went to Kentucky where his brother, Major John O'Bannon - a Revolutionary War figure - was living. Shortly after his arrival he was elected by the people of Logan County to represent them in the state legislature, where he served from 1812 through 1820.

Presley O'Bannon died on September 12, 1850 and was buried in Henry County, Kentucky. In 1919, through the efforts of the Susannah Hart Chapter of the Daughters of the American Revolution, O'Bannon's body was moved to the Frankfort Cemetery. Today many people still stop by to pay their respects to the man who, by his gallant actions, helped to "set the best traditions of the Corps." As an additional tribute, three Navy destroyers have been named *USS O'Bannon*: DD-177, DD-450, and DD-987. The latter is still on duty. The second, DD-450, was one of the most distinguished destroyers in Navy history. She served in World War II and won seventeen battle stars, the most of any destroyer, and served further in Korea and Vietnam, retiring in 1970.

55 DAYS AT PEKING

Boxer Rebellion

"The Americans who have been besieged in Peking desire to express their hearty appreciation of the courage, fidelity, and patriotism of the American Marines, to whom we so largely owe our salvation." - Arthur H. Smith and Charles E. Ewing

Shortly after the turn of the century, an allied coalition of eight nations entered a foreign land to protect the interests and safety of their peoples living in the area. Anti-Western terrorists had killed hundreds of innocent civilians and threatened to kill all "foreign devils" unless they left the region. American determination and leadership convinced the other nations to strike into the capital of the beleaguered country. British forces formed the strongest part of the alliance with the United States. After initial successes against the terrorists, the burgeoning coalition found itself fighting the army of the host state (which had turned sympathetic to the terrorists) in what could undeniably be termed a world war.

Although this may sound like a description of contemporary military activities, the above paragraph actually describes the China Relief Expedition (CRE), a short but violent international conflict which occurred over one hundred years ago.

The CRE took place in the Far East over a fifty-five day period in the sweltering summer of 1900. Popularly known as the "Boxer Rebellion," the campaign is perhaps better labeled the "Boxer Uprising," as rebellion implies a revolt against the government. In fact, the Boxers' enmity was directed not against the Chinese leaders but against the unwelcome foreign influence in their land. Most U.S. Marines are at least vaguely aware of the campaign, thanks in part to the eventual "who's who" list of American participants: two-time Medal of Honor winners Smedley D. Butler and Dan Daly; future Commandants William P. Biddle (1910–14), Wendell C. Neville (1929–30), and Benjamin H. Fuller (1930–34); and even a young civilian engineer named Herbert Hoover, who would rise to become our Nation's 31st President (1929–33). The rosters of other participating nations are similarly impressive.

Despite the event's many famous veterans (and the 1963 Charlton Heston/Ava Gardner movie *55 Days in Peking*), many Americans today might identify the Boxer Rebellion as a feud between Don King and Mike Tyson rather than recognize its long-term military and political significance.

As the first participation of the United States in a multinational coalition against a common enemy abroad, the Boxer Rebellion provides many relevant lessons for modern military theorists and diplomats alike.

The seeds for the uprising were sown gradually in the latter part of the 19th century as foreign powers expanded their presence and power in the Far East. The British wanted to continue to import their Indian opium into China. Other Western powers had similar financial interests in the area. By the late 1800s Europeans had seized Chinese ports, Christian missions were established, and in 1895 a war was lost to Japan. Not wanting to be left out, the United States sought to retain a foothold amidst the great powers' "spheres of influence" in this region. This effort was manifest in Secretary of State John Hay's "Open Door" policy proposal in September of 1899. Hay, a former Ambassador to Great Britain, expressed America's wish to prevent other powers from restricting trade by calling for open commercial access to all treaty ports and spheres of influence (and essentially endorsing the continued integrity of the Chinese state). Although the other Western powers did not officially accept the Open Door plan, unofficially they nodded in agreement - at least in Hay's eyes. The nations recognized that sharing and dividing their conquests would bring multilateral competition and tension, but knew this cooperation would also prevent unilateral exclusion.

The average Chinese peasant was quite annoyed at the growing foreign presence in the late 1800s. Certain "secret societies" fomented popular discontent by attributing a rash of floods and droughts to Chinese gods who were said to be unhappy with the "evil" Christian missionaries and their fellow converts. One such group was I Ho Ch'uan (also Yihequon), which translated variously to "The Society of

Harmonious Fists," "The Society for Unity and Righteousness," "Righteous Harmonious Fists," or "The Fists of Patriotic Union." To westerners they became simply "the Boxers." The faction had been around for many years - at least since the 1700s - but foreigners knew very little about them. The Boxers practiced martial arts and claimed that their magical, spiritual powers made them invulnerable to their enemies' swords and bullets. Wearing uniforms with red scarves, sashes and headdresses, they often carried banners into battle marked with vitriolic slogans and threats. Although they possessed some basic firearms, they were primarily armed with swords and spears - and a swelling hatred for the missionaries, their converts, and for the others "invading" their country.

Concern for the safety of Americans in China led to the stationing of U.S. Marines at the Peking (now Beijing) legation on 4 November 1898 and soon thereafter in the northern Chinese trading center of Tientsin (now Tianjin), however these guards were withdrawn in the spring of 1899 after tensions appeared to ease. The Boxers vigorously rejoined their uprising in late 1899, terrorizing rural Christian missions and their native Chinese converts in the northern provinces. The rioters burned churches and homes, raped women, and murdered families - usually by beheading and often by dismembering. Many of those who escaped fled to the foreign legations in the capital city of Peking. The Chinese Empress Dowager, Tzu Hsi (nicknamed "Old Buddha"), superficially supported the calls of concerns from foreign officials in Peking and Tientsin, but took little to no action against the Boxers. These frustrated officials wired their governments for help in May of 1900, fearing the increasing popularity and aggressiveness of the brutal Boxers operating nearby. The Boxers continued to attract potential followers by promising that millions of "spirit soldiers" would soon descend from the heavens to help rid the homeland of all foreigners. A full-blown crisis appeared imminent.

After the Spanish-American War ended in December of 1898, the Americans maintained a large military contingent in the Philippines - just four hundred miles from China - and were thus prepared to answer the distress call from China within days. Putting President William McKinley's "Manifest Destiny" platform into action, a small force of Marines and sailors landed in the port of Taku (now Dagu) on 29 May. At 10:30 that night the landing party ended their forty-mile journey at Tientsin, substantially augmenting twenty-five British Royal Marines who had been stationed there for several months. Grateful foreign residents offered the new arrivals a lively welcome, complete with brass band, free beer, and "lusty cheers for Uncle Sam."

The commanding officer of *USS Newark*, Captain Bowman H. McCalla, USN, led the U.S. naval force and accompanied a smaller detachment as it made its way northwest to Peking two days later. Thousands of silent Chinese lined the road as this detachment marched into the capital and provided an awesome spectacle for the Marines (who reciprocated by offering a similarly impressive scene for the curious onlookers), but the natives' subdued welcome betrayed the raucous terror that was to come. Under McCalla, Marine Captain John "Jack" Twiggs Myers commanded approximately fifty Marines and two sailors for the Peking guard, a unit that was followed into that city by similar contingents from Britain, Russia, Japan, France, and Italy (German and Austrian guards arrived three days later). Altogether, an international group of twenty-one officers and 429 Marines, soldiers, and sailors stood poised to defend their frightened countrymen in the Chinese capital. Fortunately, this rescue party of reinforcements arrived before the Boxers surrounded the town.

Within days the so-called Righteous Fists had dismantled railroad tracks and were severing telegraph lines, essentially cutting Peking off from the rest of the world. Captain McCalla had returned to the larger force assembled at Tientsin, and was growing frustrated at the apparent indecision and hesitancy of the assembled coalition as the situation to the north deteriorated. At an evening conference of allied commanders on 9 June, he allegedly exclaimed:

"I don't care what the rest of you do. I have one hundred and thirty men here from my ships and I'm going tomorrow morning to the rescue of my flesh and blood in Peking. I'll be damned if I'll sit down here, ninety miles away, and just wait."

The British were eager to press forward as well, and the following morning a rescue party of about two thousand men from the eight-nation coalition set off to free their comrades in besieged Peking, leaving a few hundred behind to defend Tientsin. Admiral Sir Edward Seymour, commander in chief of the British China Station, led the force, seconded by Captain McCalla.

In Peking the 473 foreign women, children and noncombatant men had consolidated inside the British legation - designed to hold only about sixty people - because it was deduced to be the most easily defended location. British Minister Sir Claude MacDonald, a respected official with previous military experience, was the obvious choice for commander of the besieged group. The senior American was the Honorable Edwin H. Conger, U.S. Minister to China. The military men manned a portion of the Tartar

Wall, a forty-five-foot high, forty-foot wide structure that bordered the northern half of Peking. While these men stood watch on the perimeter, the women made colorful sandbags from their expensive silks, curtains and dresses. Dan Daly later recalled:

"Those legation ladies were wonderful. They ripped up all their ballroom dresses to sew up sand bags for us - all kinds of colors. I never saw such fancy sand bags. Some of 'em were even trimmed with lace!"

Altogether some four thousand people from eighteen countries, including Chinese Christian converts, were crowded into the international legation area prepared to defend their lives. Requesting additional help, Conger reported to the State Department on 14 June that Peking was:

". . . in the possession of a rioting, murdering mob, with no visible effort being made by the Government in any way to restrain it. . . . In no intelligent sense can there be said to be in existence any Chinese government whatsoever."

Fortunately, additional forces from the Philippines were already on the way.

Even as the Boxers terrorized the foreigners in Peking, torching buildings and chanting "Sha! Sha!" (meaning "Kill! Kill!") by night, Seymour's troops had run into troubles of their own. Boxers pulled up the train rails in front of and behind the relief party. After several battles with the spear-wielding terrorists, Seymour's men noticed that regular Chinese Army forces - better organized and equipped than the Boxers - were also engaged in the fight. On 15 June Admiral Seymour decided that it would be impossible to reach Peking, and turned his column to initiate a fighting retrograde back toward Tientsin. That same day Boxers

13

overran the Native City of Tientsin, but the foreign concessions resisted and held. An allied naval bombardment and capture of Chinese forts at Taku on 17 June provided sufficient impetus for the Empress Dowager to drop her facade of concern for foreigners and reveal her true hostile intentions. On 19 June the Chinese Government ordered the foreign ministers in Peking to leave the country within twenty-four hours.

For their part, the Chinese probably feared the invading militaries would become occupation forces, and the Empress undoubtedly believed that, in order to maintain power as the leader of the weakening Qing (Manchu) Dynasty, she had to support the increasingly popular uprising against foreign aggression. Unfortunately, the natives elected to compound the disturbance through violence. German Minister Baron von Ketteler was murdered by a Chinese official the next morning, and the foreigners elected to "hunker down" rather than risk being slaughtered in an attempt to reach the border without protection. The following day British professor Hubert E. James was captured and decapitated, and his head was hung proudly from one of the Tartar Wall gates. China then declared "war on the world," officially backing the Boxer movement and throwing her own troops against the foreigners trapped in Tientsin and Peking - and against the surrounded Seymour expedition.

Meanwhile, on the same day that the Peking delegations were ordered to leave, the second wave of American naval forces from Cavite, Philippines arrived in China after a five-day voyage. The Marines were led by Major Littleton W.T. Waller and included eighteen-year-old 1st Lieutenant Smedley D. Butler, fresh from fighting the guerilla campaign in the Philippines. After quickly sizing up the situation the new relief force attempted to seize Tientsin but, after fierce fighting, was driven back to its starting position twelve miles from the railroad station. During the withdrawal Butler, 1st Lieutenant A.E. Hardy, and four enlisted men earned their salt by carrying a wounded Marine named Private Charles H. Carter almost seventeen miles in trace of the rear guard and in the face of constant Chinese pressure. The four enlisted Marines were awarded the Medal of Honor for this feat. Butler and Hardy were brevetted to captain.

An estimated thirty thousand Chinese fighters held positions in and around Tientsin, including Imperial soldiers with more than sixty artillery pieces and machineguns. The allied defenders, in contrast, had just 2,400 men and nine cannon to cover a five-mile perimeter. Despite this, they enjoyed excellent defensive positions, most of which had been engineered by Herbert Hoover and built by thousands of Chinese Christians. Hoover had graduated from Stanford in 1895 and married fellow geology student Lou Henry in 1899, immediately before bringing her along to China for what would become a very exciting "working honeymoon." He proved to be of great help to the defenders in Tientsin. In addition to lending his engineering expertise to the force, he "checked barricades, went on foraging raids with his young staff, [and] crept out with them at night to work the small pumping plant just outside the town," purifying drinking water from the polluted river.

Resuming the offensive, Waller's Marines and their allies successfully broke through the Chinese lines and entered Tientsin on 23 June, an action that convinced the Imperial Chinese troops to retreat to Peking. A "rescue party rescue party" then brought back Seymour's stalled expedition from their captured stronghold in the Hsi-Ku (later "French") arsenal on 26 June. Upon his return to Tientsin, a wounded

McCalla turned over official command of all American forces ashore in China to Major Waller.

Major Waller's command would be short-lived however, as additional reinforcements from the Philippines were once again on the way. American Army forces arrived in theater on 6 July, and Marine Colonel Robert L. Meade, who had arrived with eighteen other officers and three hundred enlisted men from Cavite, assumed command of the Marines and 673 soldiers from the 9th U.S. Infantry. On 12 July Meade conferred with the British leader, Brigadier A.R.F. Dorward, whose column was comprised of Bengal Lancers, Sikhs, Royal sailors and Marines, and Royal Welch Fusiliers (RWF). The commanders agreed to launch an attack to capture the remainder of Tientsin - two-thirds of which was still controlled by the Boxers - early the next morning. During the successful but costly attack, Herbert Hoover served as a guide for the Marines, and Smedley Butler was shot in the right thigh. Five British naval guns, removed from *HMS Terrible* and used earlier in the year to help defend the besieged city of Ladysmith from Boers in South Africa, provided much of the allied artillery support. Coupled with the destruction of the stores at the Hsi-ku and Tientsin arsenals and the control of the railway from Tientsin to Taku, the capture of Tientsin signaled a significant turn for the allied forces. They could now regroup and concentrate on saving those still embroiled in the action at Peking.

And embroiled they were. Eleven days earlier on 3 July, Captain Myers earned his reputation as a hero. He coordinated critical defenses on the Tartar Wall for several days in a row without pause or sleep, and after a short one-day rest returned to lead a bold night attack to secure a partially constructed Chinese tower that threatened the legation positions. Minister Conger later thanked Captain Myers, writing:

"...yours was a most trying position from the start. Our fate depended upon holding the wall. It could not have been held except for your heroic and successful charge of July 3rd..."

Years later the Royal Marines dedicated a bronze memorial in the United Kingdom that honors the Boxer conflict and highlights this particular action, accurately depicting Myers in the lead position of the assaulting American and British Marines - complementary warriors who would become true brothers-in-arms by the end of the Boxer ordeal.

After Myers' raid on 3 July, and some celebratory beer along the Tartar Wall on Independence Day, the defense consisted largely of focused sniper and counter-battery fires. Marine captain and Texan Newt Hall succeeded Myers as the Marine Corps' detachment commander, as Myers had run into a spear which had been embedded in the wall during his night raid and suffered a leg wound that soon became painfully infected.

Around this time a creative Navy gunner's mate named Mitchell fixed a discovered antique cannon and put it into good working order. The newly acquired weapon was soon put to good use. The resurrected artillery piece effectively fired improvised ammunition consisting of bags of nails and was fondly labeled "The Old Crock," and later, "Betsy."

Then on the morning of 15 July, the day after the allies to the south finally captured Tientsin, Dan Daly performed one of the many feats that would earn him the Medal of Honor for "meritorious conduct" during the siege by single-handedly manning a portion of the Tartar Wall while Captain

Hall checked on reinforcements. Effectively fighting off repeated Chinese attacks with courageous hand-to-hand combat, Private Daly heard the frustrated Chinese yell "Quon-fay" at him several times throughout the ordeal. He later learned this meant "very bad devil," an unintended but apt compliment from the enemy in a precursor to the 1918 battle at Belleau Wood where Daly would again earn America's highest award for valor, and his fellow Marines would earn the moniker "teufelheunden," or "devil dogs."

Outside of China, foreign leaders had been slow to grasp the severity of the situation. International media would help cajole their awareness as world headlines on 16 July screamed about the "massacre in Peking." Based on false rumors that the Chinese had successfully overrun the legations - killing every man, woman, and child inside - many newspapers committed what has been called "one of the most monumental mistakes in the history of journalism." Mistake notwithstanding, the German Kaiser was inspired to call up thirty thousand reinforcements, but these troops would not arrive until October - well after the relief of Peking. The Germans, however, had not been the only ones to finally grasp the scale of the growing conflict. Additional American reinforcements arrived in China on 30 July after a five-week transit from the United States.

Major William P. Biddle led almost five hundred Marines, and Army Major General Adna R. Chaffee commanded the entire American force, roughly two thousand strong. The international coalition generals chose British General Alfred Gaselee as overall commander for the allied forces and quickly organized them to relieve Peking. This final relief column departed on 4 August and was comprised of 18,000 to 19,000 men from eight nations (with almost half of the force from Japan). Ironically, after 16 July (the day of the

Peking massacre headlines), Peking enjoyed an unusual, semi-official truce until being liberated, though occasional sniper fire kept the trapped foreigners on edge and the meager supply of horse meat was running low.

Gaselee's eighty-five-mile march to Peking was punctuated by two sharp battles at Pei Tsang and Yang Tsun, however the toughest adversary during the advance was unquestionably the environment. Contributing to the environmental onslaught were scorching hot temperatures and merciless dust storms, relieved only occasionally by driving rain and mud. A dirty river contaminated with decapitated bodies and serving as the primary water source compounded the relief column's anguish. Just as the famous Chosin breakout in late 1950 arguably remains the Marines' most memorable cold weather fight, the Peking relief column's march in August 1900 China stands as an epic struggle of similar extremes, albeit at the opposite end of the thermal spectrum. In both cases the men offered exceptional performances under inhumane conditions and continuous enemy pressure. Although some of the exhausted, dehydrated men fell back and staggered into the nightly camps after dark, few American Marines dropped out of the march.

International rivalry and raw male competitiveness helped fuel the column's motivation to press forward. Unfortunately, this same competitive spirit resulted in the abandonment of plans for a coordinated attack on Peking as the Russians surprised the other allies on the night of 13 August by independently initiating an assault in an attempt to be the first to liberate the desperate legations. While the other troops were setting up their bivouacs in anticipation of a synchronized morning attack, the eager Russians pressed on without consultation or command. The other forces had no choice but to follow suit. Gaselee's coalition did succeed in capturing Peking, but to the Russian's mild disappointment it was the British and Americans who were first through the gates at midday on 14 August. The following day the allies attacked the inner "Forbidden City" (where entry was barred to all but the Imperial court), forcing the Dowager Empress to flee in disguise. The advancing liberators, however, held short before occupying the sacred inner sanctum of the capital.

Finally, after much political debate, the foreigners staged a victory parade through the never before entered Forbidden City on 28 August, essentially capping the end of the siege - although foreign forces quelled several sporadic revolts in various northern towns through October. Soldiers, sailors, and Marines from each of the eight participating nations joined in an elaborate ceremony, marching together in a grand procession that sent obvious signals to the defeated Chinese and symbolized the conviction and cooperation of the provisional allies. In a dramatic reversal, the Imperial Chinese Army rounded up and executed Boxers - fellow Chinese who had been their comrades-in-arms just days earlier. One undisputable and unfortunate wart on the campaign's finish was the disgraceful looting and pillaging of the capital and Forbidden City by foreign troops, although British and American forces, for the most part, refrained from these crimes.

By October the Marines who had participated in the Boxer Rebellion were out of China. The short event dominated world news for a few weeks during the hot summer of 1900, but public attention quickly turned elsewhere even before the Boxer Protocol was signed. The Americans were back fighting in the Philippines, and Britain would send almost half a million men to fight the Boers in South Africa through

1902. Militarily, on the surface, the Boxer Rebellion was merely another colonial "small war," but for today's student of military history this small war has much to offer. Indeed, as Professor Rick Norton remarked:

"The development of the Marines' Small Wars doctrine owed a substantive debt to the writings of Colonel (Charles E.) Callwell, the British author of 'Small Wars' fame . . . and clearly shows the continuing influence of the Victorian combat experience on military thought well into the 20th Century."

The Boxer Rebellion is a clear example of a colonial-type campaign conducted under Callwell's Victorian small war theory, but its educational value continues on several levels with valid applicability to many present-day concepts such as low-intensity conflict, joint and combined expeditionary operations, military operations on urbanized terrain, force protection/antiterrorism considerations, logistics concerns, forward presence, naval power projection ashore, and expeditionary operations.

The Boxer Rebellion also saw the birth of America's worldwide engagement and, despite its relatively minor scale, was in truth the real first world war. This war was fought before American military maneuverists had embraced the notions of "surfaces and gaps," "fog," "friction," or "centers of gravity." Nevertheless, the conflict heralded an early awakening of global American might and illustrated the value of coalition cooperation. While illustrating the capability of this rapidly comprised multinational alliance of determined partners spearheaded by American and British resolve, the crisis signaled the decline of China's two-thousand-year-old dynastic government, even as it showed their culture's willingness to respond to intrusions with military force.

The allies succeeded through the implementation of old-fashioned force-on-force attrition warfare, however it was the strategic resolve of the engaged nations' leadership - not just the tactical brashness of the battlefield victors - that ensured success. In mid-July, when victory in Peking was almost inevitable, the Chinese failed to seize the legations. This clearly was because they lacked the will and not the ability. Why then did they seek the truce on 16 July rather than exploit their advantage? Possibly they were satisfied with having taught the foreign "devils" a lesson, sensing that the allies had "lost face." More likely they knew that America and her partners would continue sending reinforcements until their mission was accomplished, and to continue the onslaught was to delay the inevitable.

The fight with the Boxers in 1900 also forged an inseparable bond of respect between the American and British Marines, and even more strongly so between Leathernecks and RWF, who repeatedly proved themselves together in combat during the campaign. The U. S. Marine Corps had not yet adopted its now-famous amphibious character, but Marine actions during the expedition helped foster concrete American public support and strong international respect for the bold, reliable, "always faithful" U.S. Marine Corps.

Arthur H. Smith and Charles E. Ewing, expressing a resolution unanimously adopted at the meeting of American missionaries in Peking on 18 August 1900, wrote:

"The Americans who have been besieged in Peking desire to express their hearty appreciation of the courage, fidelity, and patriotism of the American Marines, to whom we so largely owe our salvation. By their bravery in holding an

almost untenable position on the city wall in the face of overwhelming numbers, and in cooperating in driving the Chinese from a position of great strength, they made all foreigners in Peking their debtors, and have gained for themselves an honorable name among the heroes of their country."

All of these aspects of the Boxer campaign make it a truly exemplary and worthwhile campaign study, particularly in the chaotic times of the 21st century. The noble sacrifices and heroic deeds of Smedley Butler, Dan Daly, Jack Myers, and their fellow Boxer Rebellion veterans continue to inspire and set the standard for Marines everywhere. This high standard remains critical for American fighting men and women who will continue to prosecute our country's dispersed and challenging small wars campaign - the global war against the bona fide threat of international terrorism - well into the foreseeable future.

OLD GIMLET EYE

Major General Smedley D. Butler

"That (state) which separates its scholars from its warriors will have its thinking done by cowards, and its fighting by fools." - Thucydides, ' The Peloponnesian Wars'

Major General Smedley Darlington Butler (far right in photo), one of the most colorful officers in the Marine Corps' long history, was one of only two Marines to receive two Medals of Honor for separate acts of outstanding heroism. He was born in Pennsylvania on July 30th, 1881 and was raised a Quaker, the descendant of two old and distinguished families of Quakers. His father was Thomas S.

Butler, who was for over thirty years a Representative in Congress from the Delaware-Chester County district of Pennsylvania and a longtime chairman of the House Naval Affairs Committee. The general's mother was a Darlington and a Hicksite Friend. Though small in stature Butler was a leader of kids his age in school, although he didn't show any leaning toward a military career until the Spanish American War broke out with the sinking of the battleship *USS Maine* in Havana Harbor on February16th, 1898.

Butler was still in his teens when he was appointed a second lieutenant in the Marine Corps on 20 May 1898 for the War with Spain. Following a brief period of instruction in Washington, D.C., he served with the Marine Battalion, North Atlantic Squadron, until February 1899, when he was ordered to his home and honorably discharged on 16 February 1899.

Butler was later commissioned a first lieutenant in the Marine Corps on 8 April, 1899 and assigned to duty with the Marine Battalion at Manila, Philippine Islands. He saw some skirmishes there, and was then sent to China in June of 1900 to relieve the Foreign Legations in Peking which were under siege by a group of Chinese Nationalists called the "Boxers."

Butler served with distinction in China, and was promoted to captain by brevet for distinguished conduct and public service in the presence of the enemy near Tientsin, China. Butler landed with a force of Marines at Taku, China and was in a battle right away. They were ambushed and had to pull back. When he realized a wounded Marine was left behind Butler and five others fought their way back to the man and carried him out of harm's way, and then carried him another eighteen miles through hostile territory to a hospital. The four enlisted men in this group got the Medal of Honor, while Butler and the other officer were breveted to Captain since officers were not eligible for the Medal of Honor in those days.

After that episode he was leading a company of Marines in an attack on the walled city of Tientsin, and again carried a wounded man to safety. He himself had been shot in the leg, refused aid until all of the other wounded men were taken care of, and even then only submitted to a bandage and rejoined his men in the attack. Despite his leg wound, a fever, and an abscessed tooth, Butler exposed himself to enemy fire and dragged a British soldier back to safety. He was even struck by a bullet that glanced off one of his tunic buttons. The British army wanted to give him a medal, but he had to refuse it due to regulations forbidding a U.S soldier from getting medals from a foreign service.

In his weakened condition, Butler was shipped back to the States with Typhoid Fever in 1900. Here he was, a Marine captain, a bona fide hero of a tough military campaign, and he was yet to see his twentieth birthday. Butler, later known to thousands of Marines as "Ol' Gimlet Eye," was honored when the citizens of his native West Chester, Pennsylvania presented him with a sword on his return from the Boxer Rebellion. Some fifty years later that trophy was presented to the Marine Corps for permanent custody.

Eighteen months later he was shipped to the Island of Culebra, off Puerto Rico. It was here he was to have his first brush with the higher-ups. Butler's men were ordered to fortify the four hundred foot hills on this island, and in the heat it was back-breaking work. Any water and supplies had to be shipped in. They were then ordered to dig a canal from the ocean to the lagoon in the middle of the island. This work in the horrible heat started to affect the men, and they started dropping with exhaustion and fever.

Concerned for his men, Butler wrote to the Navy and received no help - but when his father found out about the conditions he used his influence in Washington to rectify the conditions, and the Navy was reprimanded. Butler himself never forgave the Navy brass for the way they had treated his Marines. Then in 1903 Smedley was sent to Honduras with his Marines to protect American interests from rebels, after which he was put in garrison in the States and got married in 1905. That same year he was sent again to the Philippines and again had a run-in with the Navy.

While working on gun emplacements in the hills surrounding Subic Bay, Butler's men found themselves out of supplies. Every day they saw a supply boat pass within hailing distance and tried to contact it, but the boat never acknowledged them. Finally, Butler took a native boat and some Marines and headed for the supply camp. It was a harrowing trip through a violent storm, but they made it after five hours. They then commandeered a tug, filled it with supplies and went back to their base, enduring yet another storm. The Navy was furious, sent Butler home with a "nervous breakdown," and gave him nine months to recuperate.

In September of 1908 Butler was tested and judged alright to re-enter the Corps, and a month later he was promoted to Major. In December of 1909 he was sent to Panama as Commander of the 3rd Marine Battalion as protection for the men building the canal. He was also sent at various times from Panama to Nicaragua in 1909, 1910, and 1912 to protect American interests against bandits and revolutionaries.

When the canal was finished Butler was sent to Mexico. There was a lot of anti-Americanism going on there, and American interests and citizens were in danger. He first went to Mexico City dressed in civvies to look over the situation, later landed with the Marines at Vera Cruz, and won the Medal of Honor for bravery under fire. Major Butler "was eminent and conspicuous in command of his Battalion. He exhibited courage and skill in leading his men through the action of the 22nd and in the final occupation of the city." He brushed off any praise of himself, considering it was his job to do it, and refused the medal. He continued to refuse it until a few years later when he was *ordered* to accept it, and he did. His was the attitude that he only did what any Marine would have done in his place.

The following year he was awarded a second Medal of Honor for bravery and forceful leadership as Commanding Officer of detachments of Marines and seamen of *USS Connecticut* in repulsing Caco resistance on Fort Riviere, Haiti on 17 November 1915.

During World War I, Butler commanded the 13th Regiment of Marines in France. For exceptionally meritorious service he was awarded the Army Distinguished Service Medal, the Navy Distinguished Service Medal, and the French Order of the Black Star. When he returned to the United States in 1919 he became Commanding General of the Marine Barracks, Quantico, Virginia and served in this capacity until January of 1924, when he was granted a leave of absence to accept the post of Director of Public Safety of the City of Philadelphia. In February of 1926 he assumed command of the Marine Corps Base at San Diego, California and in March of 1927 he returned to China for duty with the 3rd Marine Brigade. Then from April to October he again commanded the Marine Barracks at Quantico, and on 1 October, 1931 he was retired upon his own application after completing of thirty-three years service in the Marine Corps.

General Butler died at the Naval Hospital, Philadelphia and *USS Butler*, a destroyer which was later converted to a high speed minesweeper, was named for him in 1942. This vessel participated in the European and Pacific theaters of operations during the Second World War.

HE SERVED ON SAMAR!

The March Across Samar

"Courage is not having the strength to go on - it is going on when you don't have the strength." - Theodore Roosevelt

For a period of some two years following the cessation of hostilities with Spain, some of the wild pagan tribes of the Philippines (about five per cent of the total population of about seven million) kept the armed forces of the United States busy maintaining order. Although there had been few demonstrations by organized insurgents, the U. S. Marines in the districts of Subig and Olongapo in the Luzon Islands did good work in ridding the area of various roving bands of 'ladrones.'

But the island of Samar had for some time been a veritable hotbed of insurrection. On September 28th, 1901 the soldiers of Company C, Ninth Infantry, who were stationed at Balangiga, were massacred by insurrectos wielding bolo knives while they were in the mess hall eating dinner. It was this tragedy of Balangiga that caused Army Brigadier General Jacob M. Smith, who was in command of the military district which included the island of Samar, to call for reinforcements. That brought the U. S. Marines into action on Samar.

On October 20th, 1901 a battalion of Marines commanded by Major L. W. T. Waller was detailed at Cavite for duty on the island of Samar, the easternmost of the Visayan group, by Rear Admiral Frederick Rodgers, who was the senior U.S. Navy squadron commander of the Asiatic station. Although the Marines were placed under the command of Brigadier General Smith to reinforce and cooperate with the U. S. Army troops on Samar, it was also contemplated that Major Waller's movements should be supported as far as possible by a vessel of the fleet, to which he should make reports from time to time, and through which supplies for his battalion were to be furnished.

The battalion, composed of Companies C, D and H of the First Regiment, and Company F, Second Regiment, was equipped in heavy marching order and embarked on the Flagship *USS New York* at Cavite on October 22nd, 1901. The battalion arrived at Catbalogan, Samar on October 24, and the men and supplies were transferred to *USS Zafiro*. Preceded by *USS Frolic*, carrying Rear Admiral Rodgers and staff and Brigadier General Smith and his aides, *Zafiro* proceeded through the straits between Samar and Leyte to Tacloban, Leyte and then to Basey, Samar where Major Waller disembarked his headquarters and two companies and relieved some units of the Ninth Infantry. The remainder of the battalion took aboard a three-inch gun and a Colt automatic six-millimeter gun and proceeded to Balangiga, on the south coast of Samar, where Captain David D. Porter was left in command with 159 men and relieved the 17th U. S. Infantry, with instructions to begin operations as soon as possible. Major Waller then returned to Basey.

The area assigned to the Marines embraced the entire southern part of Samar. Active operations were immediately begun, both at Basey and Balangiga, and small expeditions were sent out almost daily to clear the country of General Vicente Lukbam's guerrillas, who usually operated in small, roving bands. The situation in the vicinity was very tense because of the Balangiga massacre and other recent happenings, hence the measures prescribed for crushing the insurrection were somewhat retaliatory. On November 5th Major Waller took a detachment to the Sohoton River and drove the guerrillas from their trenches there, with two Marines being killed. A number of small expeditions were sent up the Cadacan River and several of these parties were fired on, but the skirmishes were slight. In an engagement on November 8th at Iba, several insurgents were killed and captured. An expedition under Captain Porter, sent out to scout in the vicinity of Balangiga, killed one insurgent and captured seven, and found many relics of the massacred men of the Ninth Infantry.

As a result of the continual harassment by the Marines along the southern coast of Samar, the insurgents fell back from that region and occupied their fortified defenses on the Sohoton cliffs along the Sohoton River. About the middle of November three columns of Marines were sent into the Sohoton region to attack this stronghold, which had been reported to be practically impregnable. Two of the columns, under the command of Captains Porter and Bearss, marched on shore while the third column, commanded by Major

Waller, went up the river in boats. The plan of attack was for the three columns to unite on November 16th at the enemy's stronghold and make a combined assault.

On November 17th the shore column struck the enemy's trail and soon came upon a number of bamboo guns. One of these guns, emplaced to command the trail, had the fuse burning. Acting Corporal Harry Glenn rushed forward and pulled out the fuse. The attack of the Marines was a complete surprise, and the enemy was routed. After driving the insurgents from their positions, the Marines crossed the river and assaulted the cliff defenses. In order to reach the enemy's position the Marines had to climb the cliffs, which rose sheer from the river to the height of about two hundred feet and were honeycombed with caves which were accessed by means of bamboo ladders and narrow ledges with bamboo hand rails. Tons of rocks were suspended in cages held in position by vine cables (known as bejuco), ready to be dropped upon people and boats below. The Marines scaled the cliffs, drove the insurgents from their positions and destroyed their camps. Major Waller's detachment, coming up the river in boats, did not arrive in time for the attack - which in fact probably saved it from disaster. Instant destruction would have undoubtedly been the fate of the boats had they undertaken the ascent of the river before the shore column had dislodged the insurgents.

Further pursuit of the enemy at this time was abandoned because their rations were exhausted and the men were in bad shape. The volcanic stone had cut the men's shoes to pieces, many of them were barefoot, and all had bad feet. The Marines had overcome incredible difficulties and dangers in their heroic march, and the positions they had destroyed must have taken several years to prepare. Reports from old prisoners said they had been there years working on the defenses. No white troops had ever penetrated to these positions, and now they were held as a final rallying point.

In a communication dated December 5, 1901 Major Waller referred to General Smith's desire that the Marines make the march from Basey across the island of Samar to Hernani for the purpose of selecting a route for a telegraph wire to connect the east and west coasts. General Smith also asked Major Waller to run wires from Basey to Balangiga, and left to the major's discretion the point of departure from the east coast, either from Hernani or Lanang.

On December 8th two columns left Basey for Balangiga, with the one under command of Major Waller proceeding along the shore line, and the other under Captain Bearss marching about two miles inland. Stores were sent aboard a cutter which was kept abreast of the beach column. Although the Marines did not encounter any organized resistance, the obstacles of nature which they encountered proved far more deadly than the natives and their many contrivances. Major Waller decided to start his ill-fated march across Samar from Lanang, work up the Lanang River as far as possible, and then march to the vicinity of the Sohoton cliffs, which his Marines had recently captured.

On arriving at Lanang, Major Waller was urged not to make the attempt, however he said in his report:

"Remembering the general's (General Smith's) several talks on the subject and his evident desire to know the terrain and run wires across, coupled with my own desire for some further knowledge of the people and the nature of this heretofore impenetrable country, I decided to make the trial with fifty men and the necessary carriers."

The detachment started from Lanang on the morning of December 28, 1901 and was composed of the following personnel: Major Littleton W. T. Waller, Captain David D. Porter, Captain Hirim I. Bearss, First Lieutenant A. S. Williams, Army Second Lieutenant A. C. Lyles, (an aide sent by General Smith), Second Lieutenant Frank Halford, fifty enlisted U. S. Marines, two native scouts and thirty-three native carriers. The start was made in boats, but when Lagitao was reached it was impossible to use them further on account of the numerous rapids so the remainder of the distance was made on foot. One of the most trying features of the march was the necessity for crossing and re-crossing the swollen river many times, which kept the men's clothing wet continually.

By December 30th it was necessary to issue reduced rations, and the next day the rations had to be cut down to one-half and the number of meals per day to two. The march was continued across the rugged mountains, and on January 3rd the rapidly vanishing food supply and the serious condition of the troops made the situation critical. The men were becoming ill, their clothing was in rags, their feet were swollen and bleeding, and the trail was lost. After a conference with his officers, Major Waller decided to take Lieutenant Halford and thirteen of the men who were in the best condition and push forward as rapidly as possible and send back a relief party for the main column, which was placed under the command of Captain Porter with instructions to go slowly and follow Major Waller's trail. The advance column was afterwards joined by Captain Bearss and a corporal, the former carrying a message from Captain Porter. A message was sent back to Captain Porter, directing him to follow the advanced column to a clearing which had been found where there was a quantity of sweet potatoes, bananas and young coconut palms, and to rest there until his men were in condition to continue the march. This message did not reach Porter however, as the native by whom it was sent returned two days later stating that there were so many 'insurrectos' around that he was afraid.

On January 4th Major Waller's party rushed a shack and captured five natives, among whom were a man and a boy who stated that they knew the way to Basey. After crossing the Sohoton River, they found the famous Spanish trail leading from the Sohoton caves to the Suribao River and followed it. The party crossed the Loog River and proceeded through the valley to Banglay, on the Cadacan River. Near this point they came upon the camp which Captain Dunlap had established to await their arrival. Major Waller's group went aboard Captain Dunlap's cutter and started for Basey, where they arrived on January 6th, 1902.

Concerning the condition of the men of his party, Major Waller said:

"The men, realizing that all was over and that they were safe and once more near home, gave up. Some quietly wept, others laughed hysterically... most of them had no shoes. Cut, torn, bruised and dilapidated, they had marched without murmur for twenty-nine days."

Immediately after the arrival of the detachment at Basey a relief party was sent back to locate Captain Porter's party. The following day Major Waller joined this relief party, and remained out nine days searching for signs of Captain Porter without success. The floods were terrific, and several of the former campsites were many feet underwater. The members of the relief party began to break down due to the many hardships and the lack of food, and the group had to return to Basey, where Major Waller was taken sick with fever.

Meanwhile Captain Porter had decided to retrace the trail to Lanang and ask for a relief party to be sent out for his men, most of whom were unable to march. He chose seven Marines who were in the best condition, and with six natives set out January 3rd for Lanang. He left Lieutenant Williams in charge of the remainder of the detachment with orders to follow as the condition of the men would permit. Captain Porter's return was made under difficulties many times greater than those encountered during the march to the interior. Food was almost totally lacking, and heavy rains filled the streams making it almost impossible to follow down their banks or cross them as was so often necessary. On January 11th he reached Lanang and reported the situation to Captain Pickering, the Army Commander there. A relief expedition was organized to go after the remainder of the Marines, but it was unable to start for several days because of the swollen river. Without food, and realizing that starvation was certain if they remained in camp, Lieutenant Williams and his men slowly followed Captain Porter's trail, leaving men behind to die beside the trail one by one when it was no longer possible for them to continue. One man went insane' the native carriers became mutinous, and some of them even attacked Lieutenant Williams with bolos. After having left ten Marines to die along the trail, the surviving Marines were finally met by the relief party on the morning of January 18th and taken back to Lanang.

Lieutenant Williams, left in charge of the weakest men of the expedition, undoubtedly had the most trying task of the whole unfortunate affair. The full circumstances of his attempt to extricate these exhausted men from the midst of that wild tropical jungle is one of the most tragic yet the most heroic episodes in Marine Corps history. The entire march across Samar was about 190 miles, and Major Waller's march, including his return with the party searching for Captain Porter, was about 250 miles.

Major Waller's detachment of Marines was withdrawn from Samar and returned to Cavite on March 2nd, 1902. For many years thereafter, officers and men of the United States Marine Corps paid a traditional tribute to the indomitable courage of these Marines by rising in their presence with the following words of homage: "STAND, GENTLEMEN, HE SERVED ON SAMAR!"

QUICK AND THE DEAD

Sergeant John H. Quick

"He spelled out his message with extreme care amid the whistling snarl of Spanish bullets all 'round him, his back turned toward the enemy in apparent contempt for whatever they would do. He was magnificent." – Stephen Crane

While the U.S. Navy had prospered in the period of relative peace following the Civil War, the Army and Marine

Corps had not fared as well - the Army at that time consisted of only 25,706 enlisted soldiers under the leadership of 2,116 officers. In the face of that, military intelligence estimated that any ground force required to end the Spanish rule in Cuba would face more than eighty thousand men under General Blanco, and so just two days before the United States declared war on April 25th, 1898 President McKinley issued a call for 125,000 volunteers to train and prepare for war in the Antilles.

The call to service was met with great exuberance. People were whipped into a patriotic fervor by the stories of the yellow journalists, and it seemed everyone from aged Civil War veterans to the youngest of America's sons wanted to go help the Cubans earn their independence. From amongst these volunteers eager to fight was selected a special unit, and on May 15th former Under Secretary of the Navy Theodore Roosevelt arrived in San Antonio, Texas. He had resigned his position and received an Army commission as a Lieutenant Colonel to train and prepare the unit which was officially designated the First U.S. Volunteer Cavalry, but would forever be known as "The Rough Riders."

Ten days into the Rough Riders two-week training program at San Antonio the first Army expedition left San Francisco for duty in the Philippine Islands, and that same day President McKinley issued the call for an additional 75,000 volunteers.

At the onset of hostilities, the Marine Corps was perhaps even *more* unprepared to wage a ground war. The Corps could muster only 2,900 men, and this force was already spread thin manning fourteen shore stations from coast to coast and serving aboard forty U.S. Navy ships. Most of the Marine Corps' cadre of officers were graduates of the Naval Academy, but these leaders numbered only seventy-seven men, and several were Civil War veterans too old for field assignments.

One of these aging Civil War commanders was Lieutenant Colonel Robert W. Huntington. At the outbreak of war Colonel Commandant Charles Haywood, himself a Civil War veteran, ordered every available Marine to report for duty at New York's Brooklyn Navy Yard. Posts and receiving ships were pared to the minimum, and within days Lieutenant Colonel Huntington had mustered twenty-three officers and 623 enlisted men. They were designated as the 1st Marine Battalion, and began training immediately for combat in Cuba. Colonel Commandant Haywood took a personal interest in their progress, and ensured Huntington's Marines were well supplied for battle. The battalion's five rifle companies were issued the new Lee Navy rifles, and the artillery company was supplemented with four three-inch rapid fire guns.

During the Civil War Huntington had been a young lieutenant during the battle of Bull Run, where the savage fighting and heavy casualties taught him the importance of training. Thirty-five years later, as a Lieutenant Colonel, the wizened old warrior was determined that his young Marines would be well prepared for battle. Under his leadership they trained constantly. With thirty-five years of service and an impeccable military record, Huntington would have been a general or admiral by now if he had been in any other branch of service, but promotions were excruciatingly slow in the Corps. At that time Marines were military men admired for their unit discipline, but they were for the most part considered glorified guards for the Navy's ships and posts. Huntington's First Marine Battalion was about to change all that, and create a new tradition for the Corps.

The prevailing lack of respect for the Marines was quickly evident in the response to their presence aboard *USS Panther* en-route from New York to Key West, Florida. *Panther* had been re-christened from a former merchantman vessel to serve as transport for the First Marine Battalion. The ship's skipper, Commander George C. Reiter, saw the 623 Marines and their officers as cargo, and something of an inconvenience. When *Panther* arrived in Florida in May, Commander Reiter sought an opportunity to offload his "cargo" and requested permission to order the Marines ashore until they received further orders. When his request was granted by the commodore at Key West, Reiter quickly dumped the Marines into the swamps of the Florida Keys... without their supplies.

While that might have angered or frustrated lesser men, Lieutenant Colonel Huntington it turned into a positive. As quickly as the Marines were dumped into the swamps on May 24th, he used their situation to continue training for a war that would be fought in the similar jungles and swamps of Cuba. That impromptu training would occupy the Marines for two weeks.

Even as the Marines learned to survive the swamps, Admiral Sampson was locating the Spanish flotilla in Santiago de Cuba and was engaging in his ill-fated attempt to choke the harbor entrance with the *Merrimac* and build a Naval blockade outside that harbor's entrance. For the moment the Naval action was stalemated, and there was a concern that Admiral Cervera might be able to keep his ships protected in Santiago Harbor for weeks, perhaps even months. In the meantime Sampson's warships would have to continue their blockade of the harbor, as well as their patrols around the Caribbean island. The rotation of his vessels to

Key West to re-coal was becoming both a nuisance and a tactical bombshell.

Forty miles to the east of Santiago Harbor was Guantanamo Bay, an excellent port from which to undertake the re-coaling of the American warships. Admiral Sampson was ordered to "take possession of Guantanamo and occupy (it) as a coaling station." Upon receiving these orders, Sampson responded with a cable to both Key West and Washington, D.C. that simply said:

"The First Marine Battalion boarded the Panther on June 7th for the 3-day voyage to Guantanamo Bay."

While they were en-route, Commander Bowman McCalla took three Navy warships into the bay on a reconnaissance mission. On the morning of June 10th McCalla assembled a force of forty Marines from *USS Oregon* and *USS Marblehead* and sent them ashore to scout the area. This advance reconnaissance element made contact with local Cuban freedom fighters, scouted the proposed base area, and gathered important intelligence information. Upon returning they reported that Spanish General Felix Pareja had seven thousand troops inland and all around Guantanamo Bay.

As the morning turned into afternoon, *Panther* arrived with its 623 Marines. Even with the knowledge that they would be outnumbered ten to one, Lieutenant Colonel Huntington's leathernecks fixed bayonets and waded into the waters to make their amphibious landing. As they landed they faced their first foe on these foreign shores - but it was not the Spanish. It was the skipper of their transport ship, Commander Reiter.

As the Marines began establishing positions above the bay, their commander noted their lack of supplies. Returning to the beach he found his Marines performing stevedore duties, with their landing slowed as they were required to unload their own cargo while the crew of *Panther* simply watched in amusement. When he confronted Commander Reiter, the skipper of *Panther* once again showed his complete disdain for the men of the Marine Corps. To further infuriate Lieutenant Colonel Huntington, Reiter informed him that he had chosen to keep most of the small-arms ammunition of the First Marine Battalion aboard his ship to provide ballast.

Huntington did his best to maintain his military bearing, and immediately headed for *Marblehead* to appeal his case to Commander McCalla. A month earlier McCalla had watched as Marines from *Marblehead* and *Nashville* had entered Cienfuegos Harbor in small boats while braving rough seas, mines and point-blank enemy fire to cover the cable cutting mission. Having seen these young Marines in action, he had gained an appreciation for their courage and ability to fight.

McCalla went directly to Commander Reiter with orders that left no questions. "Sir," he bellowed to the skipper of the *Panther*, "break out immediately and land with the crew of the *Panther* fifty thousand rounds of 6-mm ammunition. In the future, do not require Colonel Huntington to break out or land his stores with members of his command. Use your own officers and men for this purpose, and supply the Commanding Officer of Marines promptly with anything he may desire."

With the inter-service rivalry firmly settled by Commander McCalla, Lieutenant Colonel Huntington and his Marines directed all their efforts to securing their positions. The unopposed landing went well, and by late afternoon the leathernecks had set up their camp. Color Sergeant Richard Silvery from C Company was the first to raise the American flag over Cuba during the war, and Huntington displayed his own respect for the commander of *Marblehead* by naming what would be the first permanent American base on foreign shores Camp McCalla.

As afternoon gave way to evening, outposts were established to protect the camp. It was inevitable that the enemy who had been a no-show for the landing would not long ignore the American presence on the island, and Huntington wanted his leathernecks well prepared. Commander McCalla promised the commander of the First Marine Battalion that, when that time came, his Marines could count on the support of naval gunfire from his fleet. McCalla further demonstrated his respect for Huntington by telling him, "If you're killed, I'll come and get your dead body."

The Marines had twenty-four hours of unmolested opportunity to establish their presence on foreign shores, and then the enemy came. On June 11th, Company D was attacked by a Spanish force. Under the leadership of their company commander, Lieutenant Wendell C. Neville (a future Commandant whose heroism sixteen years later at Vera Cruz, Mexico would earn him the Medal of Honor), the leathernecks did their best to repulse the initial probe. The first shots attracted the attention of all the Marines, as well as the bevy of reporters who had followed them into Guantanamo Bay to write stories for their newspapers back home. One of them wrote:

"Up from the sea came a line of naked men, grabbing their carbines and falling into place as Lieutenant Colonel R. W. Huntington issued his orders getting a formation in a semicircle behind the brow of the hill, and waiting to see how much force would develop against them."

The untested leathernecks of the First Marine Battalion responded to their training. Most had spent the day stripped to underwear in the tropical heat, and with the first sounds of gunfire rallied to meet the enemy. The same correspondent continued:

"There was no fun in this for naked men, but they held their places and charged with the others."

Much of the history of the 'Splendid Little War' was preserved for future generations because of the competition among newspapers and magazines back in the United States for readers. Throughout the war, seldom did a force move without a large contingent of correspondents. Indeed, as naval ships moved from place to place, even while landing troops, movements were often hampered by the criss-crossing of smaller boats carrying the eager reporters. The media presence included some of the best-known names in American publishing, including the combat art of famed western artist Frederick Remington. Also joining the Marines at Guantanamo Bay was the now-famous young author of *The Red Badge of Courage*, Stephen Crane. Crane was reporting for *McClure's* magazine, while nearby *Moby Dick* author Herman Melville also was observing the leatherneck's operation and filing dispatches for the news at home.

That first battle was brief, a quick hit-and-run of the American defenses. The prompt response by the Marines and shelling from *Marblehead* soon caused the 'Dons,' as the Spanish were called, to pull back. In the quiet that followed the Marines assessed their casualties and found two, Privates William Dumphey and James McColgan of Company D. Huntington quickly ordered Captain George Elliot and Company C to pursue and find the enemy. Meanwhile,

Huntington himself led another patrol along with Captain Charles McCawley and Sergeant Major Henry Good.

The patrols fought their way through the tropical foliage, quickly learning the nuances of jungle warfare. Before Elliot's leathernecks could locate the enemy, the enemy found *them*. The Dons had used the foliage to their advantage, hiding their presence until the patrol was almost upon them, and then sprung their ambush. Fortunately the same heavy jungle that provided camouflage also made accurate fire difficult, and none of Elliot's Marines were seriously wounded.

Huntington's frustration came not so much from the days and nights of fighting as from the manner of combat. The Spanish soldiers would hide in the jungle, creep into places of concealment to snipe at unsuspecting Marines, and then quickly fade back into the foliage. It was almost like fighting ghosts, only these ghosts were systematically wounding or killing young American Marines with relative impunity. About the only positive thing to happen for him in four days came on Monday morning, when more than fifty insurgent Cuban soldiers arrived with their leader, a Cuban Colonel.

During the day the weary Marine Battalion Commander had ample opportunity to visit with these newly arrived Cuban soldiers, men who knew the terrain well and had a good grasp of the enemy force encamped in and around Guantanamo Bay. Despite the fact that General Blanco had at least eighty thousand soldiers throughout Cuba, and in spite of the earlier report from a forty-man Marine recon unit that as many as seven thousand enemy occupied the hills and jungles in the extreme southeast part of the island, the Cuban rebels estimated that there were only five hundred to eight hundred troops in the immediate area.

As the two military commanders watched the Marines suffering in the sweltering sun, pausing occasionally to sip the somewhat tepid water that filled their canteens, the Cuban colonel reminded Lieutenant Colonel Huntington that maintaining a fresh supply of water was a challenge for all military men on the island - Americans, Cuban freedom fighters, and even the Spanish. The turn of the conversation, combined with the new information about enemy strength in the area, slowly lead to a bold and daring plan to break the stalemate.

Into the evening Lieutenant Colonel Huntington laid out his plans, briefing his officers and NCOs while conferring further with the Cuban colonel. He had learned from the Cubans that the Spanish soldiers in the area got their own drinking water from a well at Cuzco, about three miles to the east of Camp McCalla. On the following morning Huntington would dispatch a contingent of his Marines, prepared for war in the jungles, to defeat the Spanish defenses at Cuzco and destroy their well. It was at once both a small measure of revenge, and a solid tactical effort that would make war in the jungle much more difficult for the enemy.

At dawn on Tuesday, June 14th the Marines of Companies C and D moved out of Camp McCalla towards Cuzco. The 150-man assault force was reinforced by the fifty Cuban rebels and, as they began their trek eastward, *USS Dolphin* began a slow steam parallel to the men along the coastline. Under the command of Captain W.F. Spicer, the three-mile trek was doubled as the Marines slowly wound their way along the jungle trails and over the hills.

Stephen Crane accompanied the men on their mission, watching events unfold around him with a reporter's eye and later transcribing them with the same colorful language that had made *The Red Badge of Courage* such powerful reading:

"The Marines made their strong faces businesslike and soldierly," he reported. "Contrary to the Cubans, the bronze faces of the Americans were not stolid at all. One could note the prevalence of a curious expression - something dreamy, the symbol of minds striving to tear aside the screen of the future and perhaps expose the ambush of death. It was not fear in the least. It was simply a moment in the lives of men who have staked themselves and come to wonder which wins - red, or black."

The men of the First Marine Battalion were making history which, when subsequently reported in the flowery language of Crane and Melville, would make the exploits of these leathernecks the precursor of the Marine Corps of the future. Their amphibious assault, being the first combat troops in hostile territory, and now an offensive against the enemy, would provide heroes for the reading public at home and inspiration to thousands of future Marines on foreign shores.

Three days of sleepless nights and constant danger began taking its toll on the Americans, and the hot sun combined with the arduous trek began to quickly sap any remaining strength. Nearly halfway to Cuzco, these factors began to take its toll on the force. Several Marines began to suffer heat stroke. Faces flushed, minds becoming numbed and disoriented, and cramps setting in, several had reached the limit of their endurance. Among those to fall victim was the commander, Captain Spicer.

Halfway to their destination Company C commander Captain George Elliott assumed command of the force from the ailing Spicer. In the distance *Dolphin* cruised just offshore, and a stretcher party was mounted to move down to the beach with the casualties of the heat and jungle conditions. Then the assault force continued its march to Cuzco.

A little over two miles from their destination Elliott commanded First Lieutenant L.C. Lucas to take his platoon, along with half of the Cuban rebels, and flank the advancing men of the main assault force. Lucas' men moved out with the intention of surprising any enemy pickets between the assault force and Cuzco, and of cutting them off from the fortifications about to come under attack. Hot, sweaty, tired and moving forward through sheer guts and determination, sound discipline began to falter as the leathernecks forsook the painstaking, slow movement through the jungle. As they stumbled ever forward the enemy outposts quickly noted their presence, and withdrew to the protections at Cuzco. By the time Captain Elliott's main assault force reached its destination the element of surprise was gone and the Spanish garrison was armed and awaiting the Americans arrival.

Six companies of riflemen of the Sixth Barcelona Regiment manned the gun ports at Cuzco as the Marines arrived. Elliott's quick recon revealed a large, horseshoe-shaped hill nearly a thousand yards from the enemy. The high ground dominated the landscape and would provide the Marines with a tactical advantage should they be able to reach it. Elliott gave the command, and his Marines began the frantic rush to its crest. Enemy gunfire erupted as the weary leathernecks ignored their exhaustion and the heat as they pushed their bodies beyond reason, and at that great distance the heavy Mausers of the enemy were unable to unleash a lethal torrent of fire. Their rounds "sang in the air until one thought that a good hand with a lacrosse stick could have bagged many," wrote Stephen Crane.

Marine Private Frank Keeler was less flowery but more succinct when he penned his observations in his diary:

"Up the hill we charged in the face of fire, but we drove them back in disarray."

As the leathernecks scrambled for the heights they paused only long enough to return fire, and First Lieutenant Neville began to shout orders across the hilltop as he rallied his men. The boom of his voice became one of the most memorable events of the day, leading to his Marines bestow upon him a nickname. It was a moniker he would carry with him in the years to follow. From his actions sixteen years later in Mexico which earned him the Medal of Honor, to his years as Commandant of the Marine Corps, he would be facetiously but affectionately remembered as "Whispering Buck."

The sounds of the battle at Cuzco could be heard all the way back to Camp McCalla, and Lieutenant Colonel Huntington quickly dispatched Second Lieutenant Louis Magill and fifty men from Company C to cut off any Spanish withdrawal. A second contingent under First Lieutenant J. E. Mahoney was also dispatched to reinforce Captain Elliott's assault force.

Meanwhile, Captain Elliott requested heavy fire support from the port guns of *USS Dolphin*, which was still pacing the Marines just off the Cuban beaches. On the high ridge, the leathernecks whooped with glee as one of the warship's big shells slammed into a blockhouse below, sending frantic Spanish soldiers fleeing in all directions. The accurate long-range fire of the Marines, combined with the heavy shells from *Dolphin*, had a devastating effect on the enemy. Quickly the battle began turning into a rout, but amid the cheers of the Marines disaster loomed.

A short distance away Lieutenant Magill's fifty-man force was furiously engaging the retreating Dons as the shells from *Dolphin* mopped up the action. Suddenly, as the shells fell on the Spanish, they also began raining deadly missiles on Magill's leathernecks. From the heights, Captain Elliott's Marines were about to helplessly witness the annihilation of their comrades by friendly fire.

Sergeant John Quick suddenly grabbed his blue polka-dot bandanna and quickly attached it to a stick. From the high hill he could see *Dolphin* in the distance, and surely in his exposed position they would be able to see him as well. So too, could the Dons. Crane wrote his own account of what happened next:

"I watched his face, and it was grave and serene as a man writing in his library. I saw Quick betray only one sign of emotion. As he swung the clumsy flag to and fro, an end of it caught on a cactus pillar. He looked annoyed."

The only way to effectively signal the American ship was to ensure that they could see him. As the enemy rounds whistled through the air all around him, he bent to his task. Crane continued:

"He spelled out his message with extreme care amid the whistling snarl of Spanish bullets all round him, his back turned toward the enemy in apparent contempt for whatever they would do. He was magnificent."

Sergeant Quick's use of the flag to advise *Dolphin* via semaphore code resulted in the immediate end to the deadly rain of naval gunfire. Lieutenant Magill and his fifty Marines were spared death by the quick thinking and intrepid action of a lone, leatherneck sergeant.

Within an hour it was all over, with the Spanish soldiers who survived the battle at Cuzco pulling back into the jungle and retreating. Elements of Captain Elliott's strike force gave pursuit, while others entered Cuzco and destroyed the well shortly after three in the afternoon. Casualties for the Marines had been light, with a total of one wounded and twenty heat casualties. Two of the Cuban insurgents were killed, and two others wounded. Losses for the Spanish were much higher, though hard to estimate. Eighteen enemy soldiers were captured, perhaps as many as sixty killed, and as many as one-hundred fifty wounded. Thirty enemy Mausers were captured as well.

At home Senator Henry Cabot Lodge wrote of the Marine triumph at Cuzco:

"The Marines had done their work most admirably and fought with the steadiness and marksmanship of experienced brush fighters."

For his heroism at Cuzco Sergeant John Quick was awarded the Medal of Honor, along with a young Marine Private named John Fitzgerald. Lieutenant Colonel Huntington's First Marine Battalion had done their job so well, and fought so fiercely, that they gave the enemy cause to thereafter avoid them. Not only had their battle at Cuzco Well destroyed the enemy's water supply, it had robbed them of their will to fight.

While Huntington returned his valiant Marine force, which by now had grown to nearly half of the full 647-man battalion, to Camp McCalla, Spanish survivors straggled back to the city of Guantanamo to advise General Pareja that they had been attacked by ten thousand Americans - and that disturbing news caused the enemy general to halt all attacks at Camp McCalla. And so, with the exception of those

Marines stationed aboard Navy vessels, the Spanish-American War was a brief but bitter four-day affair that rewrote Marine Corps history.

BOIS DE LA BRIGADE DE MARINE

Belleau Wood

"What shall I say of the gallantry with which these Marines fought? I cannot write of their splendid gallantry without tears coming to my eyes."
– Major General James Harbord, U.S. Army

Belleau looks, at first glance, like a thousand other small sleepy country towns in the Champagne region of France. There is a small square with a little pond full of goldfish, a pale yellow mailbox, an extremely modern phone booth, and a small bakery which doubles as social center and general store. The houses are solid old farm buildings, built lovingly by earlier generations with the heavy gray fieldstones of the region. Occasionally a car moves through town, while women in their black headscarves stop and chat with each other as they go about shopping.

But in what *other* ordinary village in France would you find the Marine Corps' colors proudly displayed behind the mayor's desk, next to the French Tricolore. Where else in France do people smile radiantly when they spot the USMC sticker on your car - they even stop and chat with you, trying out their few words of English. "Ah, les Marines," they say, "our friends, nos amis." In only a few American towns would U. S. Marines find so kind a reception as they have been finding in the village of Belleau for the past few decades. This unusual friendship between U. S. Marines and the people of Belleau was formed in the last summer of the First World War.

When dawn broke on June 6th, 1918 pounding artillery and mortar fire, the staccato of heavy machineguns, the flat cracking sounds of Springfield '03 rifles, the howling, the cries of the wounded, and shouts of encouragement broke the country quiet. It was then that American Marines placed themselves firmly into the collective memory of the people of Belleau. These ties are exceptionally strong and have held firm over several decades, through another World War, and much political upheaval - and will, if the Marines and the people of Belleau get their way, last for a long time to come. For most people here and in Europe, the First World War is ancient history. Few know the names of individual battles such as the Battle of Belleau Wood which "saved Paris" in the words of French Prime Minister Georges Clemenceau. But for Marines and the people of Belleau, the memory is still strong.

Claude Crapart raises Charolet cattle. This type of cattle is one of the most prized in France for its beef. In June of 1981, close to Belleau, one of his cows stepped on an unexploded 75mm artillery round left over from that battle in 1918, and died. Claude Crapart also produces wheat, corn, sugar beets and eggs. He is also the mayor of Belleau, and lives on an old farm in the center of town. You still can see traces of machinegun fire and shrapnel on some of the walls.

"I hope," he said some time ago, "that Marines will always find a warm welcome in Belleau, that our little village will be for them a little bit of their own place, for they came to this country when it was in danger. I think I know my fellow villagers well enough to know that they share this hope with me, particularly the older ones. Those who saw the battlefield with the horrified eyes of a three, ten, or fifteen-year old child tell us intimately about it."

The mayor of Belleau was speaking to an audience that included virtually all the villagers and some twenty Marines. Among the Marines was Colonel James L. Cooper, who had presented the mayor with a set of Marine Corps colors and an invitation to visit Washington, D.C.

Marine Corps colonels don't just travel around handing out flags and inviting mayors of small country towns to Washington. Colonels, in relatively short supply, have better things to do, and the Marine Corps is too concerned with saving money to do something of this nature in the first place. However, for Belleau, the Marine Corps will do a lot, including paying the way to Washington for the mayor of this village of eighty souls, honoring him with a parade, and showing him around Marine installations.

In May of 1918 the German Army embarked on a major offensive to end the war. Superbly led and disciplined, even after four years of demoralizing trench warfare, the German troops poured through an opening in the front and advanced to within fifty-six miles of Paris. A panic started. Parisians packed their belongings and started to flee. French troops, ordered to make a stand, retreated in disorder. Senior Allied generals and key French politicians seriously considered asking for an immediate armistice. The French government prepared to flee. The situation was quite desperate. To delay the rapidly advancing Germans, the Allied high command ordered the American 2nd Division to make a stand. That division, fresh to the war, included the 4th Brigade, which consisted of the 5th and 6th Regiments of Marines and the 6th Marine Machinegun Battalion.

The Marines moved up to the front. An Army officer described them:

"They looked fine, coming in there, tall fellows, healthy and fit – they looked hard and competent. We watched you going in, through those tired little Frenchmen, and we all felt better. We knew something was going to happen."

The Marines passed disoriented, demoralized, thoroughly beaten and scared French troops. A high-ranking French officer instructed the Marines to move back to safety. Resistance was useless. "Retreat? Hell, we just got here!" Marine Captain Lloyd Williams retorted. The French commander in charge of the sector that the Marines were to defend told 2nd Division commander Brigadier General J. G. Harbord, "Have your men prepare entrenchments some hundreds of yards to rearward." Harbord is reported to have replied, "We dig no trenches to fall back on. The Marines will hold where they stand." Up to that time the two Marine regiments had barely been tested in battle. Most of the men had just arrived in France, after some brief training at Parris Island and Quantico. One could easily dismiss their youthful enthusiasm, but over the next weeks they showed that they meant every word of it.

On June 4th, 1918 the 2nd Division held some eleven miles of front just fifty miles east of Paris. Astride the old Paris-Metz highway, the Americans faced crack German regiments. The Marines just north of the highway were up against the 461st Imperial German Infantry Regiment.

The Germans advanced. From an incredible distance of eight hundred yards, Marines took the Germans under fire. As their ranks thinned, the Germans retreated and decided to dig in. They made their stand in a wooded area known as *Bois de Belleau,* a game preserve long favored by the Kings of France and named for a small nearby village. Official German reports from the battlefield recount the amazement German officers expressed over the marksmanship of the Marines. The Springfield '03 rifle, carried by Marines, was not supposed to be accurate for much more than 450 yards. Marines were hitting their targets at almost twice that, an incredible distance. In today's Corps, the maximum distance shot for rifle qualification is five hundred yards - but marksmanship has always been a hallmark of Marines.

The Marines, their spirits high after that first success, were ordered to take the wood and expel some twelve hundred Germans. On June 6th they faced a wide, rolling wheat field, behind which was the hilly area of the woods. Just four feet high, the wheat offered little cover to the charging Marines from heavy German machinegun fire. Artillery and mortars supported the Marines, and the Germans had the same.

Three times the Marines charged. Three times they fell back. It was the hottest day of the year, so the Marines shed their blouses and attacked with nude torsos. Water ran short. There were many casualties. Only after the fourth charge across the field, which remains a wheat field to this day, did they succeed in gaining a toehold in the woods. *Chicago Tribune* reporter Floyd Gibbons (who lost an eye at Belleau) wrote:

"The oats and wheat in the open field were waving and snapping off – not from the wind, but from the rifle and

machinegun fire of German veterans in their well-concealed positions. The sergeant (Dan Daly) swung his bayoneted rifle over his head with a forward sweep. He yelled at his men, 'Come on, you sons-of-bitches. Do you want to live forever?'"

A German report after the battle read:

"The Second American Division must be considered a very good one, and may perhaps even be reckoned as storm troops. The different attacks on Belleau were carried out with bravery and dash. The morale effects of our gunfire cannot seriously impede the advance of the American riflemen."

"I am up front and entering Belleau Wood with the U.S. Marines..." reporter Gibbons wrote on the evening of June 6th. When his story appeared in the United States, it ignited an explosion of publicity about Marines which did much to build their reputation. Reports from the battlefield show that virtually no man who entered the woods was without some wound. The casualties on that day were incredible. There were some sixteen hundred - close to twenty percent of the brigade's strength.

S.L.A. Marshall, the famous military historian, wrote:

"Here at Belleau Wood, the German commander was to risk all 'in a local dogfight.' And he had picked on the wrong people."

The battle would go on for another twenty days, and by June 26th the woods had been emptied of Germans. Major Maurice Shearer, commanding officer of the 3rd Battalion, 5th Marines, sent his famous message: "Woods now U.S. Marine Corps entirely..."

The French were relieved that the German advance had been stopped, and to honor the Marines they officially renamed the woods "Bois de la Brigade de Marine," a name that sticks to this day.

The casualties had been major. During the twenty days of battle the 4th Brigade lost 1,062 dead and 3,615 wounded, about fifty-eight percent of brigade strength. It's a small figure measured against the millions lost in the battles at Verdun and the Argonne, but it was an enormous figure for the Corps - which at the time numbered just 75,000 men.

Emile Richard is an old man. He raises some cattle, a few chickens, and a few pigs. Emile lives alone on top of a small hill near Belleau with his three dogs, his lovingly polished hunting rifles, and his memories. He was six years old when the Marines occupied his father's farm on 5 June, 1918. The farm still stands, although part of it was never rebuilt after being demolished by German artillery. He remembers the sounds of battle, the wounded, the dead. He remembers a young Marine lieutenant, Victor Bleasdale, drawing a map on his kitchen table, and setting up machinegun positions for the attack. You still can see the shallow pits around Emile Richard's farm. Even the map is still there. It was left behind, and Emile now proudly displays it to visitors. It's his proudest possession. Not much happened in his life. There was the battle of Belleau and a few years ago the young lieutenant who became a general returned to visit him. Today in his home is an old crucifix. Jesus lost one hand and his right arm in this long ago battle.

One sultry spring evening, over a couple of beers at the Paris Marine House in the U.S. Embassy, the Marines talked about their frequent visits to Belleau. They go there for Memorial Day parades, spend weekends with the Boy Scouts there, and one Marine even got married in the village church.

Some just drive out to be alone and enjoy the woods. "I just get lost, spellbound, when I go there," said one Marine. "I think about what it would have been like if I had fought there. In boot camp you have classes all the time about the history of the Corps. They tell you about Belleau Wood and they tell you about the fighting there and you take pride in that and all of the tradition. But when I finally got to Belleau, it was different. A part of my history is there since I am a Marine, and Marines fought and died there. I walk through the fields into the woods and out to the perimeter where the Germans had their machinegun nests. It's part of *my* history. I enjoy it. Like most Marines, when we talk about the Corps we bitch about it... but deep down you've got your pride."

Some years ago several young lieutenants visited Belleau while on liberty. They wore their Marine uniforms. "They were quite well behaved," recalls James Neill, who for years maintained the Belleau cemetery. "They walked around and looked at things. Then they came to the wheat field where so many Marines died that day. The wheat was close to being ready for harvest. Those lieutenants lost all their composure. One started, the others followed. In full uniform they charged across the field, yellin' and screaming, fallin' down in the dirt, crawling as if under fire, scrambling up again, charging the silent woods. It was quite a scene. The farmer whose field they demolished was quite upset... the next day. But when he heard that Marines did it, he was actually pleased. Strange people, those Marines, those people of Belleau."

Every May the village of Belleau celebrates American Memorial Day. Marines come from Paris and parade up and down through the village. There is a little ceremony, some speeches, music, and normally a detachment of French Marines. When it's all over, they assemble in the courtyard of the old castle of Belleau (which was destroyed in the First World War) and drink champagne. "They treat you like kings out there, if you are a Marine," said a Marine at the U.S. Embassy in Paris. All the village turns out for the party. Cookies are served, and much champagne. Normally they run out of glasses. Marines and villagers share the same glass, toasting a battle of long ago.

Few tourists find their way to Belleau. That's all right with the people of Belleau, and with the Marines. "It's our little secret," says an old woman sweeping a courtyard.

There are many traces of the battle. There is the small village church, rebuilt by the Americans after the war. It has stained glass windows. One shows a Marine along with a French Poilu. There is also a small plaque which reads, "For the children of Belleau who died 1914-1918."

There is a plain grave in the village cemetery. A man named Ernest Stricker is buried there. The villagers say he is a former Marine. He came back to Belleau after the First World War and saw all the graves of his friends, and one night he went out and committed suicide. "He wanted to be with them," says an old man. The people of Belleau got together, bought a plot and buried him. His grave is being maintained to this day. "He had no family," says the old man.

The military cemetery at the edge of the woods is a pleasant, serene place. It is colorful, with large massifs of forsythia, laurel, boxwood, Japanese plum, deutzia, mock orange, Oregon grape, and beds of polyanthus. There are 2,288 Americans buried there, about half of them Marines. There are also the graves of 249 unknowns. During the Second World War the German Army left the two American flags flying, even after the United States entered the war. An honor guard was even posted by the Germans - their way of paying respects to the heroism of the Marines.

The German cemetery is some 350 yards away. Both cemeteries border on the wheat field which proved fatal for so many. The dates on the headstones are the same: June 6, 1918... June 8, 1918... June 26, 1918. On the graves in both cemeteries are many flowers.

In the woods proper, close to the sign with the official name "Bois de la Brigade de Marine," is a black granite and steel monument bearing a life-size bronze bas-relief of a Marine, his torso bare, attacking with rifle and bayonet.

There is also an abundance of old artillery pieces, heavy guns, and light mortars. On some you still see the bullet holes. Others have been spiked by the Marines, their barrels split wide open, like the gaping mouth of a shark. Those guns were drawn by horses and oxen. Many of those animals died in the battle. "They had gas masks for the horses," says a Frenchman, "but not for the oxen."

Marines earned their nickname, "Devil Dogs," at Belleau Wood. When German troops first occupied Belleau in the beginning of the war, they found the courtyard of the old castle guarded by ferocious bull mastiffs. The Germans called those the "Hounds of Belleau" and "Devil Dogs," and later referred to the Marines by the same name.

The Hounds of Belleau had lived for centuries in the old castle, guarding the property, and occasionally going out on a hunt. They were massive dogs, some weighing as much as two hundred pounds. Today, in the ruins of the old castle, the life-size bronze head of a bull mastiff spews ice cold, crystal clear water. Marines, after the battle of Belleau, are said to have quenched their thirst at this fountain.

After the battle France awarded the 5th and 6th Marines the Croix de Guerre, and to this day these regiments wear the green and red fourragere in recognition of their Croix de Guerre awards. Franklin D. Roosevelt, Assistant Secretary of the Navy during World War I, was responsible for another legacy of Belleau Wood. While inspecting the 4th Brigade in August of 1918, the future President directed that enlisted Marines be allowed to wear the Corps' emblem on their collars, as officers did.

There are more traces of Belleau in today's Marine Corps. Though not quite as obvious, they may be more important. Rifle squads and fire teams, the smallest organized units in combat, remain to this day the key to Marine Corps infantry tactics. Battle-tested in three wars, and occasionally modified, rifle squads and fire teams were first used at Belleau. On June 6th, 1918 the Marines did at first what they had been taught. They attacked in a close formation, and ran up against heavy German machinegun fire. It did not work. The Marines tried it three times, and casualties mounted. Jury-rigging what would become future Marine Corps doctrine, they split up into small teams. While one team charged, another would cover it with rifle and machinegun fire. The Germans, unaccustomed to such novel tactics, failed to respond in time and by nightfall the Marines had secured a toehold in the woods. "The Marines went back to Indian warfare," said a Marine officer who studied the battle. "They did not play by the rules of the First World War. Those rules were 'when you meet resistance, you dig trenches.' The Marines did not dig trenches. They used trees and rocks for cover, attacked in small groups, and concentrated leadership on the individual Marine. It worked."

It's hard to imagine today's Corps without this battle. In a sense, the Corps proved itself to be a viable land fighting force. The 4th Brigade was the largest Marine force ever fielded. Up until then, Marines had been fighting only in company-size or smaller units, often against poorly trained or equipped enemies. For the first time, Marines fought against a professional army that had been victorious over crack English and French troops. Up to that time many officers in the other services, especially in the U.S. Army, doubted that those "soldiers of the sea" could fight a regular land battle. The Marines proved they could. Major General John A. Lejeune, who was to become Marine Commandant, was even appointed to command the 2nd Division during the Meuse-Argonne offensive.

The memory of the battle lingers in the minds of the old men of Belleau who sit down in the early evening to drink the light beer of the region. The hospitality is overwhelming. Elsewhere in France outsiders are regarded with suspicion, or ignored altogether. Here the people greet you, and take time to talk to you. As the mayor says, "a little bit of their own place." Agrees a Marine in Paris, "it's like coming home at times."

But you have to walk through the woods. Traces of the battle are everywhere. There are eleven tons of metal in every acre. Nobody can build there. Unexploded shells, some containing high explosives, others mustard gas, contaminate the ground. Every once in a while the soil turns up a shell. A cow will step on it and die. Every spring more metal surfaces. Old helmets, old canteens. Once in a while human bones are found.

But it is the trees that tell the story. Those trees that saw the battle as young, fresh saplings still carry the scars. Some have been so mutilated by shrapnel that one wonders how they survived. Bark hides rusty shrapnel once destined for a man, but caught by a tree. Occasionally the grounds keeper will cut down one of these survivors. Live ordnance was even found *inside* one tree - it grew old carrying its lethal bits of metal. But when the last survivor of the battle is dead, the remaining trees will still display their ancient scars.

DO YOU WANT TO LIVE FOREVER?

Sergeant Major Daniel J. Daly

"I'd rather be an outstanding sergeant than just another officer." – Dan Daly

Daniel J. Daly was born on November 11th, 1873 (a day late, as it turned out...) at Glen Cove, Long Island in New York, and enlisted in the Corps on January 10th, 1899 at the age of twenty-five. His professed reason for enlisting was to participate in the Spanish American War, however soon after completing boot camp he was transferred to the Asiatic Fleet.

On the evening of August 14th, 1900 then-Private Daly and Captain N.H. Hall occupied a barricade in the city of Peking, China during the Boxer Rebellion. Set between the Ch'ien Men and Hata Men gate, it was a solid defensive position.

As night fell the Captain returned to get reinforcements, and Daly volunteered to stay at the barricade. His position was assaulted by the Chinese all through the night, but the lone Marine held firm through attack after attack.

On December 11th, 1901 Private Dan Daly was awarded the Navy issue Medal of Honor. The citation for his first of two awards of the Nation's highest decoration reads, "In the presence of the enemy during the battle of Peking, China, 14 August 1900, Daly distinguished himself by meritorious conduct."

Daly's next action saw him at Vera Cruz during the Mexican American War in 1914. This was followed smartly by action in Haiti during the first occupation of that Caribbean country. By now a Gunnery Sergeant, Daly was part of a patrol which was pushing the bandit Cacos into an old French fort in an attempt to consolidate and destroy the remaining rebels.

His patrol of thirty-five Marines was ambushed by a force of approximately four hundred Cacos while fording a river. Although all the Marines made it to the riverbank safely, the horse carrying the unit's only machine gun was killed and abandoned in mid-river along with many others. During the night the embattled Marines were again attacked, and the patrol leader called for the machine gun. Daly immediately volunteered to return to the river and retrieve the weapon.

After making his way back to the river through enemy patrols he found the dead horse, cut the gun from it and, while under heavy fire strapped it to his back and returned to the Marine position. This action earned him his second Navy issue of the Medal of Honor, and a place in Marine Corps history shared by only one other Marine - Smedley Butler. Both men earned their second award during the same action. Daly's citation reads:

"Serving with the Fifteenth Company of Marines on 22 October 1915, Gunnery Sergeant Daly was one of the company to leave Fort Liberte, Haiti, for a six day reconnaissance. After dark on the evening of 24 October, while crossing the river in a deep ravine, the detachment was suddenly fired upon from three sides by about 400 Cacos concealed in bushes about 100 yards from the fort. The Marine detachment fought its way forward to a good position, which it maintained during the night, although subjected to a continuous fire from the Cacos. At daybreak the Marines in three squads advanced in three directions, surprising and scattering the Cacos in all directions. Gunnery Sergeant Daly fought with exceptional gallantry against heavy odds throughout this action."

By now Daly was forty-four years old and looking to the clouds of war in France, and was soon shipped "over the pond" as First Sergeant of the 73rd Machine Gun Company. His many actions during this conflict were to net him his, as he said, "a hat full of medals." One was earned by wiping out German machine gun nests alone using grenades and a .45 Colt pistol, and another by capturing thirteen enemy soldiers.

At Lucy li Boucage, on the outskirts of Belleau Wood France, Daly made a comment which still thunders with the

Marine spirit today. Outnumbered, outgunned, and pinned in a poor position, the Marines were soon to be chopped to pieces by the German machine gunners. Daly ordered an attack, leaping forward and yelling to his men, "Come on you sons of bitches! Do you want to live forever?" He and his small group of Marines then surged out of their position and captured the town of Lucy li Bocage.

Daly remained single his entire life, retired from the Corps on February 6th, 1929 as a Sergeant Major, and died peacefully at Glenade, Long Island on April 28, 1937 at the age of sixty-five.

SEND MORE JAPS!

Wake Island

"I've got some bad news, and some good news. The bad news is we will be filling sandbags until the next Japanese attack. The good news is we have plenty of sand." – Marine Gunnery Sergeant on Wake

Air raid sirens are wailing in San Francisco. The Army reports a flight of thirty Japanese planes over the city, heading inland. People living along the coast of Oregon and

Washington expect an invasion fleet off their shore at any moment. Up and down the Pacific coast and as far inland as Boise, Idaho cities are blacked out. Crowds gather on street corners and yell at motorists who are driving with their lights on.

The Pacific Telephone and Telegraph Company building in San Francisco is barricaded with sandbags. Some thousand homes around Monterey and Carmel are evacuated. It is decided to move the Rose Bowl game, which is scheduled for New Year's Day, 1942, to the Duke University stadium in North Carolina. The winter racing season at Santa Anita is canceled.

In Washington, D.C. the War Department is just as jittery and unsure as the average citizen. Excited officers cluster in the corridors and speculate on where the next attack will come. "Anywhere" is the answer. No one knows where the enemy fleet that attacked Pearl Harbor is. The locks of the Panama Canal could be the next target, or the aircraft factories in California. A high government official, in a state of hysteria, telephones the White House to demand that the entire West Coast be abandoned and a defense line set up in the Rocky Mountains.

It is Tuesday night, the 9th of December, 1941. Millions of Americans sit glumly and tight-lipped around their radio sets to hear the familiar voice of President Franklin Delano Roosevelt as he confirms their worst fears. It was just a little more than forty-eight hours ago that they were jolted into war.

"So far, the news has all been bad," Roosevelt says.

Rumors, confusion, uncertainty and anger have gripped the nation. The lines at the recruiting offices have been blocks long during the day. Many of them are remaining open tonight to handle the surge of volunteers. Some young men got in line before dawn to be the first to enlist - to fight back, to take revenge.

"We have suffered a serious setback in Hawaii," Roosevelt continues.

No one knows with any certainty, not even the president himself, how serious things are - only that the situation is bad and likely to get worse. Where will the Japs strike next?

At Pearl Harbor in the Hawaiian Islands military forces brace themselves as best they can, with what is left, for another attack. They expect not just a bombing raid, but a full-scale invasion of Oahu or one of the other islands.

Reports are coming in from Wake Island - it is being bombed. A dispatch from the Reuters News Agency out of Shanghai says that Wake has already been occupied by the Japanese. Why would the enemy want Wake? The generals and admirals know the enemy expects to use it as a springboard for attacks on Midway and Hawaii. In his broadcast, Roosevelt says solemnly that the news from Wake is confused, "but we must be prepared for the announcement that (it has been) seized."

The attack on Pearl Harbor is the most humiliating and overwhelming military defeat the United States has ever suffered. The entire Pacific Fleet has been wiped out, or so it is thought at the time. In the "little Pearl Harbor" attack in the Philippines a day later, MacArthur's air force is almost totally destroyed on the ground and only a few planes remain.

Everywhere in the Pacific in the days and the weeks ahead America and her new allies, the British and the Dutch, will reel from one lightning blow after another. Everywhere there will be losses and retreats and defeats, one humiliation after another. Disasters and surrenders will become daily occurrences, except on one tiny, lonely isolated island that few people have ever heard of. The early reports are wrong. Wake Island has *not* been taken. It is holding its own. The Marines are fighting back! For sixteen days following Pearl Harbor it will be bombed every day but one, and still the Marines will hold.

"Marines keep Wake!" reports the *New York Times* on December 12th. Wake Island becomes a watchword, a symbol, a heroic stand on the part of a handful of the glamorous leathernecks. Their courage eases some of the hurt and shame of Pearl Harbor. The *Washington Post* says that Wake Island has become "the stage for an epic in American military history, one of those gallant stands such as led Texans 105 years ago to cry 'Remember the Alamo!'"

"Wake Up!" becomes a national slogan. Newspapers trumpet the story of the siege of Wake on the front page in edition after edition. And then, on December 11th, an incredible event takes place. The small garrison fights off an invasion attempt, sinks two enemy ships, and damages others. These are the first Japanese ships sunk by American forces.

It is America's first victory, the first small step on what everyone knows will be a long and painful road back. Americans stand a little taller that day and swell with pride at the cockiness of the Marines. A new phrase enters the consciousness of the nation. When the garrison is asked if it needs anything, the reply is, "Send us more Japs!" This quickly becomes a rallying cry, a gallant flippant message that captures and reflects America's determination to fight back and win.

At a press conference on December 13th, President Roosevelt echoes the pride of 130 million Americans as he pays tribute to the Marines of Wake. Yet he tempers his words with a note of caution. He knows Wake may not be able to hold out much longer. The public should be prepared for the possibility of another defeat. Hopes must not be allowed to rise too high. On the same day, the *New York Times* says in an editorial:

"They have held the fort and kept Old Glory flying... they may yet be annihilated... but they have fought gallantly, and by their gallantry they have carried on in the finest tradition of their Corps."

And still they continue to hold. Casualties mount, defenses are shattered, and Japanese bombers return day after day. In great secrecy, a task force is assembled amidst the ruins of Pearl Harbor. Troops and supplies are loaded aboard ships, and the small fleet glides out into the unknown waters of the Pacific on a course that will take them two thousand miles westward. "We're headed for Wake!" the Marines shout. The island's defenders are told they can expect reinforcements.

The attrition continues while the ships move cautiously closer and closer. Then, 425 miles from Wake - less than a day's sailing time - they turn away and head back to Pearl Harbor. On the aircraft carrier *Saratoga*, Marine pilots curse their leaders. Some break down in tears. The lives of fellow Marines are at stake. On the bridge of the flagship, officers urge Admiral Frank Jack Fletcher to disobey his orders and continue on to Wake. Angry words are exchanged. Some consider the situation close to mutiny. An admiral walks off the bridge in bitterness.

And from Wake comes the message: "Enemy on island - issue in doubt." But there is no doubt. There can be no beating back this new invasion, not without reinforcements. The Marines are too few and too tired and too poorly equipped.

But for sixteen days they had given America hope while the world held its breath.

Adapted from the book *Wake Island* by Duane Schultz

MARINE'S MARINE

Lieutenant General Lewis B. "Chesty" Puller

"A battalion commander didn't need a staff, he got out in front of his battalion and he led it." - Chesty Puller

Chesty Puller is one of the Corps' best-known and most-loved characters. All of those words – "best-known," "most-loved," and "character" - fit him particularly well. In our history the Marine Corps has had at least its share, and probably more than its share, of colorful characters - but none more colorful than Chesty. When Brigadier General Edwin H. Simmons was a twenty-one-year-old lieutenant and a platoon leader in the Officers' Candidates Course in 1942 he saw him - and heard him - for the first time. The battle for Guadalcanal was just ending, and a number of officers were being brought back to the Marine Corps Schools in Quantico to talk about it. Chesty Puller was one of those who were brought back. He had commanded the 1st

Battalion, 7th Marines, and on the night of 24 October his battalion had held a mile-long front against the efforts of a Japanese regiment which was seeking to break through to Henderson Field. For this action Puller had been awarded his third Navy Cross, an award for valor second only to the Medal of Honor.

Simmons remembers sitting in the auditorium of Breckinridge Hall and seeing him for the first time on the platform. Not very tall, he stood with a kind of stiffness with his chest thrown out, hence the nickname "Chesty." His face was yellow-brown from the sun and atabrine, the anti-malaria drug that was used then. His face looked, as someone has said, as though it were carved out of teak wood. There was a lantern jaw, a mouth like the proverbial steel trap, and small piercing eyes that drilled right through you and never seemed to blink.

Puller was then forty-four years old. He had been born in the village of West Point, Virginia, where the waters of the Pamunkey and Mattaponi rivers come together to form the York River. That's in the area known as Tidewater, and he never lost his Tidewater, Virginia accent. His full name was Lewis (pronounced "Lewie") Burwell Puller and he was named for an ancestor, Lewis Burwell, who had settled in Virginia in the mid-17th Century. His father was a wholesale grocery salesman. His grandfather, Captain John W. Puller, had ridden with Jeb Stuart in the Civil War and had been killed at Kelly's Ford in 1863. Young Lewis grew up on stories of the Confederacy. He finished high school in 1917 and then went to Virginia Military Institute for a year, but when June 1918 came he told the Commandant of Cadets that he was leaving because he didn't want the war in Europe to end without him. He enlisted in the Marine Corps, went through boot camp at Parris Island, and stayed on as a drill instructor. To his chagrin, this kept him from going to France.

After the war had ended Puller was sent to Officers' Training School, commissioned a second lieutenant in the Marine Corps Reserve, and placed on the inactive list. This was in June 1919. He then re-enlisted in the Marine Corps and was sent to Haiti where he served as a lieutenant in the gendarmerie. He remained there for five years, was in forty engagements against Haitian bandits, and began to build a reputation as a bush fighter.

Puller came back to the States in 1924 and received a commission as a second lieutenant in the regular Marine Corps. He spent two years in the States, including an unsuccessful try at flight training at Pensacola, two years at Pearl Harbor, and then in 1928 he went to Nicaragua. There he served as a captain in the Guardia Nacional. (The Gendarmerie d'Haiti and the Guardia Nacional of Nicaragua were both native constabularies officered for the most part by Marine Corps officers and NCOs serving at one or more levels above their Marine Corps ranks.)

Puller won his first Navy Cross for a series of five patrol actions fought against followers of one Augusto Cesar Sandino, a name that has since reappeared in the news. During the recent unrest in Nicaragua the opposition to the Somoza government took onto themselves the title "Sandinista." As a matter of interest, President Tachito Somoza, himself a West Point graduate, was the son of Tacho Somoza, who was a Marine-trained member of the Guardia and who came to power after the Marines departed Nicaragua in 1933.

Chesty Puller came back to the States in 1931, went to the Company Officers Course at Fort Benning, and returned again to Nicaragua. One of the Managua newspapers welcomed him back with a front-page bulletin that read, "Los Marinos traen al 'Tigre de las Segovias' para combatir a Sandino." Translated, it means, "Marines bring back the Tiger of Segovia to fight Sandino."

Sandino is supposed to have welcomed Puller back by placing a price of five thousand pesos on his head. Puller, still a first lieutenant in the Marine Corps, went back into the Guardia as a captain and was soon back in his old haunts in northern Nicaragua, where he soon won a second Navy Cross for his patrol work.

The Marines never did catch Sandino, but after they left the country Somoza, who had become Jefe Director of the Guardia, caught him in 1934 and had him shot. By that time Puller had gone on to North China for service with the Legation Guard in Peiping. Here, amongst other things, he commanded the famous Horse Marines - a detachment from the Legation Guard mounted on Mongolian ponies. A year later he went aboard the cruiser *Augusta*, flagship of the Asiatic squadron, as commanding officer of the Marine Detachment. He came back to the States in 1936, and was assigned to The Basic School at the Philadelphia Navy Yard as a drill and tactics instructor. Puller saw three classes pass through The Basic School, and left a well-remembered mark on those young lieutenants who would be the captains, majors, and lieutenant colonels of the Marine Corps in World War II. On parade, his leather and buttons seemed to shine more brightly than anyone else's, his khakis were more stiffly starched, and the creases in his trousers more sharply edged. Also, and more fundamentally, he had not only his own experiences in Haiti and Nicaragua to recount, but also was deeply read in military history.

In 1939 Puller left Philadelphia and went once again to China to serve a year afloat on *Augusta* after which he joined the 4th Marine Regiment in Shanghai. He stayed with the spit-and-polish 4th Marines until August of 1941, four months before Pearl Harbor. It was a good thing that he left when he did, because the 4th Marines left Shanghai at the end of November to go to the Philippines and were surrendered to the Japanese by an Army General named Wainwright when Corregidor fell.

Back in the States, Puller was given command of the 1st Battalion, 7th Marines in the newly formed 1st Marine Division - and this is the battalion that he took to Guadalcanal. While he was there the Marines were visited by Eleanor Roosevelt, who was pretty tough herself in her own way. On meeting Puller, she asked him, "Tell me, Colonel, what are you fighting for?" thinking no doubt of some lofty set of ideals. Puller thought for a moment, and then answered, "For five hundred and forty-nine dollars a month, ma'am."

During one question-and-answer period he was asked his opinion of Marine Corps staff action at the battalion level. He growled back that a battalion commander didn't *need* a staff - he got out in front of his battalion and he led it. And that was his tactical philosophy - Chesty led from out front and he expected his officers to do likewise, which perhaps accounts for the high rate of casualties amongst his officers.

Puller spoke not only at Quantico, but also at many Army installations - Fort Benning, Fort Sill, Fort Riley, Fort Ord, and Leavenworth to name some of his stops. His message was that one American, properly trained, could handle two Japanese. Marine Raiders were then very much in the news. Carlson's Raiders had landed at Makin Island, and Edson's Raiders had fought well at Tulagi. One of the questions put to Puller at Quantico asked his opinion of the Raiders.

"Marine Raiders," he rasped, "are just Marine riflemen with extra privileges."

He was happy when the lecture tour was finished and he was sent back to the 1st Marine Division. The tour had gotten him a personal letter of appreciation from General George C. Marshall, Chief of Staff of the Army, but when Puller got back to his old regiment in April of 1943 and was assigned as Executive Officer, he wrote the Commandant of the Marine Corps:

"It is respectfully requested that my present assignment to a combat unit be extended until the downfall of the Japanese government."

The 1st Marine Division's next operation was Cape Gloucester, beginning in December of 1943. Puller landed with the 7th Marines and personally led two battalions which had lost their commanding officers in the successful assault of a heavily fortified Japanese position. For this he received his fourth Navy Cross. In February of 1944, still at Cape Gloucester, he was given command of the 1st Marine Regiment, and Puller, now a full colonel, would command the 1st Marines in the bitter fight for Peleliu in September and October of 1944. Following his philosophy of leading from the front, his regiment took terrible casualties - fifty-six per cent of the regiment were dead or wounded in nine days of fighting. The 1st Marines, over Puller's protest, were pulled out of the line before the battle was finished and sent back to their base in the Russell Islands. Chesty received a Legion of Merit for Peleliu, but did not again command combat troops in World War II.

When the Korean War began in June of 1950 Puller was commanding officer of the Marine Barracks, Pearl Harbor. It was a very pleasant post indeed. With him were his wife,

Virginia, who was from Saluda, Virginia and whom he had married in 1937 after a long courtship, and his three young children. The oldest child, Virginia McCandlish, had been born in Shanghai in 1940. The twins, Martha and Lewis B. Puller, Jr., were born in 1945. But Puller was not one to sit out a war in comfort.

He went to the cable office and bombarded the Commandant, Assistant Commandant, and the Commanding General of the 1st Marine Division with messages asking for an assignment. He was given command of his old regiment, the 1st Marines, which had gone inactive after World War II and was being re-formed at Camp Pendleton.

While the unit was in Japan getting ready for the Inchon landing, all of the field grade officers - that is, majors and above - in the regiment were gathered onboard the amphibious command ship *Appalachian* to be briefed on the forthcoming operation. The intelligence officer described the objective area, told them about the twenty-foot tides and the sea wall, and said that they would be making an evening landing which would give them only about an hour to consolidate their positions. The operations officer then went into detail on the objectives. At that point the gathered officers were looking rather apprehensively at each other, and it was then that Chesty got up and planted himself in front of them.

"Forget what the S-2 told you," he growled. "We'll find out what's on the beach when we get there. You people are lucky. In my day we had to wait twenty years for a war. You get one every five years. You people have been living by the sword. You better, by God, be ready to die by the sword."

With those words of encouragement, they went back to their battalions. A couple of days later, on September 17th, the regiment was firmly ashore at Inchon. The 3rd Battalion had passed into reserve and was bivouacked next to the 1st Marines' command post when word began to spread that General MacArthur had come ashore and wanted to give Chesty a medal. Chesty sent back a message. "If he wants to decorate me, he'll have to come up here."

And that's exactly what MacArthur did. He arrived by jeep, complete with crushed hat, sunglasses, corncob pipe, and leather jacket - and escorted by a body guard consisting of about a platoon of soldiers and two platoons of newspapermen and photographers. MacArthur presented Puller with a Silver Star, or actually, the promise of a Silver Star, because he had been rather liberal that morning and his aide had run out of medals. In due course Puller would also get a second Legion of Merit from the Marine Corps for the landing at Inchon, the crossing of the Han River, and the capture of the city of Seoul.

Having taken the city, the Marines provided security for the triumphant entry of MacArthur and President Syngman Rhee into Korea's capital. After Seoul, the 1st Marine Division re-embarked, moved around to the east coast of Korea, landed at Wonsan, moved north to Hungnam, and then further north and inland, away from the sea, to the Chosin Reservoir. It was there that the Chinese hit them. The 5th and 7th Marines were out in front near a place called Yudam-ni. The 1st Marines had its battalions strung out in three positions to the rear with ten to fifteen miles between them. The 3rd Battalion was located at Hagaru-ri, the 2nd Battalion and regimental headquarters were at Koto-ri, and the 1st Battalion was at Chinhung-ni.

A Chinese division hit the 3rd at Hagaru-ri. They managed to hold them off until joined by the 5th and 7th Marines, who had to fight their way through three Chinese divisions to get there. The unit made it back to Koto-ri on the 7th of

December, by which time the Division had made contact with a total of eight Chinese divisions. It should be pointed out that the Marines were fighting not only the Chinese, but also the weather. There was snow and temperatures as low as twenty-five degrees below zero. The regiment was barely inside the Koto-ri perimeter when the senior officers were summoned to the regiment's operations tent. The scene was different, but in some ways the dialogue was the same as it had been three months earlier on *Appalachian*. They were told that the 3rd Battalion would be the rear guard from Koto-ri to Chinhung-ni - an honor that they would probably have been just as happy to let someone else have. The S-2 painted a gloomy picture of how many Chinese stood between the Marines and Chinhung-ni. The S-3 then outlined a complicated scheme of maneuver under which the battalion would occupy a succession of positions on the high ground on both sides of the road. Then, after all the Division's transport had passed, the rear guard was to come out of the hills and fall in on the rear of the column. Once again the officer's faces must have shown their apprehension, because Colonel Puller, who had not yet spoken, got up, braced himself with one foot up on a ration box, pulled his pipe from between his teeth, and growled.

"I don't give a blank-blank how many blank-blank Chinese laundrymen stand between us and the blank-blank sea, they don't stand a blank-blank chance of stopping a U. S. Marine Corps regiment." You can fill in those blanks with words of your own choice. Having said this, Puller paused and said quietly, "And Christ, in His infinite mercy, will see you safely through."

The 1st Marines moved out on the 9th of December and had hard fighting for the next several days. Because their artillery was also moving to the rear they had very little artillery support and had to depend on their own mortars to give them the high-angle fire they needed to get at the Chinese in defiladed positions. One of the 81mm mortar sections was firing in support of Company H - it was called it How Company in those days, it would be Hotel Company now - and was running out of ammunition. They had left Koto-ri with two truckloads of 81mm ammunition, but one had broken through the ice crossing a stream and had to be blown up, and the other had gotten lost in the Division column. In desperation the platoon commander sent his section sergeant down to the road with instructions to stop all trucks that seemed to have ammunition, and to hi-jack any 81mm mortar ammunition they might have.

The sergeant did exactly what he was told, and Chesty Puller came roaring up in his jeep to see who was holding up the column. The sergeant told him what he was doing, and Chesty said, "I'll give you a hand." And there he was, in the middle of the road, stopping trucks, and when they found one with 81mm ammunition, helping to off-load the cloverleaves.

For the break-out from Koto-ri to Hungnam, Chesty received his fifth Navy Cross. No other Marine has ever won five Navy Crosses. For the same action he also received the Army's Distinguished Service Cross. In January, after the 1st Marines had evacuated Hungnam and were back in camp at Masan in the south of Korea, Puller was promoted to brigadier general and made the Assistant Division Commander.

General Puller returned to the States in May of 1951 and immediately attracted national attention - not all of it favorable. In San Francisco he told the press:

"What the American people want to do is fight a war without getting hurt. You can't do that anymore than you can go into a barroom brawl without getting hurt." He also said, "Unless the American people are willing to send their sons out to fight an aggressor, there's just not going to be any United States. A bunch of foreign soldiers will take over."

He was also quoted, not quite accurately, that troops should be given beer and whiskey instead of ice cream and candy. This caused a furor in the press, and a blizzard of protesting letters descended on Headquarters Marine Corps. Puller was then sent to his next duty station, Camp Pendleton, to take command of the new 3rd Marine Brigade - with the admonition to make fewer speeches.

Chesty was promoted to major general in 1953, and in the following year was given command of the 2nd Marine Division at Camp Lejeune. There he ended his legendary career. And to this day, as recruits at Parris Island and San Diego prepare to retire for the evening, they shout out in tribute, "Goodnight Chesty Puller, wherever you are!"

This chapter was adapted from a speech given by Brigadier General Edwin H. Simmons, USMC (Ret.) at the Fairfax Optimist Club Luncheon on 9 November, 1970.

GUNG HO

Makin Island

"Find a way, or make one." – General of Carthage Hannibal Barca

In the middle of the 1930s the command of the Marine Corps recognized the danger of Japanese aggression and sought ways of combating it in the future. One of the problems for the Marines was going to be the conduct of operations on many small islands where the Japanese could be expected to have intimidated or befriended the natives. Under such circumstances the Marines would have to use special tactics, and they were considering all the options. Out of this came the recognition in Washington that it would be useful to have special forces of highly trained men who could slip ashore on Japanese-held islands, conduct a raid, and come back with information to guide the Navy in its operations. And so the idea for the Marine Raider battalions was born.

High in the hills overlooking San Diego, the men of the 2nd Raider Battalion trained. From Camp Elliott they set out on two thirty-five mile hikes each week, and one seventy-mile overnight hike in which they bivouacked in the open. On the other days of the week life was easier. They only had a ten-mile hike, plus training in unarmed combat, weapons and marksmanship. Their survival course included judo, the use of the trench knife, throwing knives, boxing, stalking, and silent movement through the bush. They were given more training in the use of the bayonet. They learned to use the new plastic explosives to blow bridges and rail lines. They practiced going without food and living on very little water. They learned how to find water, and how to conceal themselves in open country.

Lieutenant Colonel Evans Carlson was their example. He led the longest hikes, and his pack was always the heaviest in the battalion. He stood in the chow line with the men, and he made up his own rack in his quarters. He kept his belongings in his footlocker, and it was always ready for inspection like any other Marine's.

Carlson had joined the Marine Corps as a private at the end of World War I, and had worked his way up through the ranks. He served with the 4th Marine Regiment in Shanghai, and the Legation guard unit in Peking. In 1938 he applied for permission to spend some time with the Chinese Communist 8th Route Army in North China, and permission was granted by both the Marine Corps and the Chinese communists. Carlson spent several weeks observing the Chinese during one of their campaigns of harassment against superior Japanese forces, and decided that a specially trained Marine unit could use similar tactics to operate successfully behind enemy lines. It was also here that he learned the phrase, and the concept, of "gung-ho," or "working together."

After several weeks of training, the battalion went down to the shore. All Raiders had to learn to swim and pass swimming tests. They also had to learn to manage a rubber boat in darkness. Once the Marines had become familiar with the procedures for doing so and had gone through some drills, they were taken out into the middle of the bay and put into their boats. Their destination was San Clemente Island, and their mission was to get ashore unobserved and make their way to a rendezvous point.

Once the Raiders had perfected their boat skills they were supplied with two submarines in order to conduct special training operations, and after a few lessons they gave an exhibition for Admiral Chester Nimitz. The Admiral and some of his staff had come down to the beach on western Oahu to witness a night landing operation, and were quite disappointed when they did not see anything. Apparently Carlson and his Raiders had goofed up. But then, just as Nimitz and his staff were preparing to return to base, Raiders appeared all around them. They had come up to within fifty yards of the staff officers before they were detected. Had it been an enemy outpost, it would have been wiped out. That was the convincer. Nimitz decided the Raiders were ready to go, and his staff had a job for them.

In July of 1942 the Joint Chiefs of Staff had encouraged Admiral Nimitz to begin preparation for an attack on the Gilbert Islands as part of the "road back across the Pacific" campaign, and less than three weeks later Lieutenant Colonel Carlson's 2nd Raider Battalion was ready to go into action in the first stage of that plan.

The Gilbert Islands had been captured from Britain at the outset of the war, and no one really knew what the Japanese had done with them since that time. The atoll consisted of several islands, but only two were of any importance - Makin and Tarawa. Makin, the larger island, was considered to be the most important because it housed a seaplane base. Admiral Ray Spruance, who was Nimitz' chief of staff, wanted to know what was on the islands and how stout the defenses were in order to plan for an invasion.

Using aerial photographs which showed the Japanese installations in detail, engineers built a mock-up of Makin Atoll at Barber's Point on the southwest corner of Oahu, and using their two submarines, the Raiders practiced conducting landings there. After much practice throughout the month of July, every man could find those positions in the dark.

As the beginning of August approached Carlson had only one major worry about the operation. He had ordered outboard motors which would be used to propel the rubber boats into the shore at Makin, and after they arrived and had been tested in training he discovered that this brand of motor had a tendency to fail when swamped with water because it had no motor cover. He protested to the supple people, but was told there was no time to bring in a new order. He would have to live with the outboard motors he had.

The submarines *Argonaut* and *Nautilus* lay in the Pearl Harbor submarine base on August 8th, loaded up with supplies and rubber boats and the thirteen officers and 208 enlisted men of the 2nd Raider Battalion. The submarines were big boats, nearly four hundred feet long, but they were not built to carry passengers. As a result, the interior was frightfully crowded. Most of the torpedoes were taken ashore to make room for more men, and extra bunks were crammed into every nook and cranny. When the hatches were battened down, the temperature inside the hull rose to above ninety degrees.

They sailed that day, destination Makin, mission reconnaissance. Most of the voyage was made submerged, with the men lying on their bunks because there was no place else to go. They had ten minutes on deck twice a day, once in the morning and once at night. That lasted for the first eight days of the voyage. But when they came to a point where they were within Japanese air range, the luxury of the few minutes on deck had to be withdrawn on the morning watch.

Nautilus arrived off Makin at three o'clock on the morning of August 16th and waited for the other submarine at the designated rendezvous point. The weather turned foul, and the commanding officer of the submarine began to worry about the possibility of his boat sliding onto the reef. *Argonaut* arrived later in the day, and at 9:15 that night the two submarines surfaced and headed in toward the Makin shore. The landing point was reached at 2:30 AM on the morning on the17th, but the weather was unfavorable, with squalls and strong winds. A decision had to be made, and Carlson did not hesitate. He gave the order, and his men began inflating their rubber boats and putting them over the side. Then they affixed the outboard motors that would propel them to the shore of Butaritari Island - but just as Carlson had feared, the motors began to take on water and conk out. The men would have to paddle the boats to shore.

The Raiders had been split into two groups, with one company aboard each submarine. Lieutenant Plumley's 'A' Company was to land on one beach, and Captain Coyte's 'B' Company on another. They would then complete their assigned missions, return to their respective beaches, and get back to the submarines by 9 PM - but with the high seas, the loss of the outboard motors, and a serious difficulty in communication, Carlson decided to land both companies at the same beach. Once the Marines were away the submarines moved out to a point four miles off the reef where the water was deep and awaited developments.

In the confusion caused by the darkness and bad weather one boat commander, Lieutenant Peatross, did not get the word about the change in plans. His boatload of twelve

Marines moved out according to the old plan and landed on its original beach. The other seventeen boats headed stealthily toward Carlson's beach to carry out the new plan, and came in through the surf without incident. So far the mission was going almost according to plan.

At 5 AM the Marines were ashore and began concealing their boats with camouflage netting. There was no sign of the Japanese - the Marines had achieved the surprise they had wanted. The fifty-man Japanese garrison, however, was on alert, because the 2nd Marine Division had landed at Guadalcanal the week before and the Japanese were watching for more landings. In any case the Japanese, who were dug in at the lagoon on the opposite side of the atoll, did not know the Raiders had come ashore.

All went well until just before six o'clock, when one Marine accidentally fired his rifle. The sound seemed as loud as a sixteen-inch battleship gun to Carlson. He knew the shot had to have been heard somewhere, and that the element of surprise was lost. Carlson immediately ordered his men to move out on their assigned missions, because it was important for them to get going before the Japanese found them. No one knew how many Japanese were on the island or what weapons they might have.

Carlson sent Lieutenant Plumley across the island to the lagoon side where there was a road, with instructions to set up a roadblock and stop transportation. As they moved, so did the Japanese, who had heard the stray shot as Carlson had feared they would. As a result the Raiders moved out in skirmish style, showing a ragged line which would not give the enemy a natural target. In the meantime Company A remained just off the beach and waited to see what would happen.

As the Americans moved across the island they were met by a band of native Gilbert Islanders, who were friendly. They told the Marines that the Japanese troops were concentrated at Ukiangong Point Lake and On Chong's wharf, which was the landing place for ships. Carlson radioed *Nautilus* with that information, and she laid down a barrage of twenty-four shells with her three-inch deck gun.

Company A's progress across the island was soon slowed by Japanese machinegun fire, but they soon made it to the road and began moving. When the Raiders spotted the first machinegun they attacked and killed the enemy gunners. The Japanese then staged a 'banzai charge.' Shouting and running, they came at the Americans with rifles and swords. The Marine fire teams opened up with their rifles, tommy guns and BARs. The superior firepower stopped the Japanese charge, and left more than a dozen men dead in the sand.

From the natives the Marines learned there were two Japanese ships about five miles off the island which were heading out to sea to avoid the invading troops. One was a 3,500-ton inter-island cargo vessel, and the other a patrol vessel of about 1,500 tons. Carlson passed the word to the two submarines, and they tried to open fire on them. The problem was, the gunners could not see their targets - so Lieutenant Plumley got into a position where he could observe and helped direct the fire.

But even with Plumley's help the submarine fire was ineffective until two men were sent ashore from *Nautilus* to spot for the gunners. They eventually got the range and put several shells into the cargo ship, setting her afire. The crew and about sixty Japanese special landing troops, who were roughly the equivalent of American Marines, abandoned ship and began to come ashore. The submarines then diverted their fire to the patrol vessel, and soon she was dead in the water too.

Plumley's men then moved toward the settlement. They captured Government House, which was the old British seat of government. They then took Government Wharf - but then the Japanese held. Plumley reported that the defenders had four machineguns and two mortars, as well as a flamethrower and some automatic rifles. There were also a number of snipers concealed in the trees.

The firing became intense. One Marine stood up to move, and before he took a step he was hit in the right arm and shoulder by five .25-caliber bullets. Sergeant Thomason moved along the line, encouraging the men.

"Gung-ho, Raiders!" he was shouting.

The Japanese mounted another banzai charge, but once again the superior Raider firepower made the difference. The charge was stopped by heavy fire from Garand rifles, Browning automatic rifles, and Thompson submachine guns.

Sergeant Thomason was then hit, and fell. It was fire from a sniper, who could not be found up in the palm trees. The firing continued sporadically. Plumley's men then learned how to deal with the snipers. The Japanese had never heard of the Garand semi-automatic rifle, because according to their manuals the Americans were armed with the 1903 Springfield bolt-action rifle. After a man pulled the trigger of a Springfield, he would have to eject the spent shell with the bolt action and then throw another round into the chamber - which would give a sniper a few seconds to fire a shot. But with the Garand, a number of rounds could be fired in rapid succession. The Americans would fire a single shot, and when the Japanese would stick up their heads the Marines would blast them.

The Japanese staged their third banzai charge, and once more they were repulsed. Afterward the Japanese commander radioed his headquarters that his men were dying gloriously in battle... but, they were dying.

Help for the Japanese was already on the way. Just after ten-thirty in the morning a pair of Japanese reconnaissance planes appeared to investigate. When they arrived it forced to submarines to crash-dive, and that took their naval guns out of the land battle. The Marines continued to move forward, but the going was harder without artillery support.

Back on the beach at his command post, Lieutenant Colonel Carlson's radioman tuned in on 'Radio Makin,' which was still operating. They knew then it would not be long before planes from Jaluit and Mille, island groups about 250 miles away, would be coming over and making things even more difficult.

Carlson was growing nervous. Anything could happen. Any sort of reinforcement might occur. The Raiders were taking too long. He paced the beach, "smoking that stinking pipe of his," as one Raider put it. He then put down the pipe and chain-smoked cigarettes. Finally he went up to the perimeter to see what Plumley and his men were doing.

As the line of advance slowed Carlson ordered a platoon from Company B to reinforce Plumley, and Lieutenant Griffith led his men forward. The Japanese fired steadily and called out insults. "Roosevelt bastards," they shouted, not knowing how right they were. Major James Roosevelt, the son of the President, was right there with Carlson.

"Marine, you die," shouted the Japanese, but it was the Japanese who were dying for the most part as the Raiders pushed slowly ahead into the former British colonial settlement.

The Japanese reconnaissance planes circled lazily, and then circled again. Then each of them dropped a single bomb in the area where the Marines were fighting and moved off to the north. Carlson knew they were headed for base to report and get help.

About an hour and a half later, shortly after 1:30, Carlson heard and then saw a dozen Japanese planes headed for the atoll. Four of them were Zero fighters, and the rest were a mixed bag of bombers including two big Kawanishi flying boats. For the next hour and fifteen minutes they zoomed in on the island, dropping bombs and strafing American positions.

Soon Carlson learned the bombing and strafing was just the beginning. The big flying boats had another mission, one which was even more dangerous to the Marines. They were bringing in reinforcements. One plane landed off King's Pier, and thirty-five Japanese soldiers waded ashore. The flying boat was brought under fire by the Marines, who disabled it so it could not take off. That meant the loss of another half-dozen reinforcements for the garrison. Carlson then moved a .55-caliber antitank gun down to the beach near King's Pier. It began firing on the Japanese aircraft as they came in, and shot down one seaplane.

Carlson had wondered what had happened to Lieutenant Peatross' boat, until an exhausted Marine came to the command post to tell the story. Peatross had not gotten the word about the change in orders, and his twelve men had landed at 7 AM about a mile from the point they had been assigned to take. They quickly encountered the Japanese, and set fire to an enemy truck. Then they got into a firefight and killed eight Japanese, but lost three men doing it. The Japanese were now between Peatross' men and Carlson's force.

The mission was supposed to last only one day, and that day was nearly over. The Raiders had not accomplished much. But what *could* they accomplish? Just now they were under heavy sniper fire, and the advance into the government area had been stopped by the reinforcements. Carlson moved his men back a hundred yards, and the Japanese moved up to occupy the positions they had just vacated. Just then another flight of bombers came over. They must have been instructed by the previous group, because they bombed where the Americans had been an hour earlier and rained their bombs down on Japanese troops.

Further south on the island Lieutenant Peatross was trying to move up and make contact with the Carlson force, but the Japanese had him stopped. He then thought about moving out independently, and getting back to the beach and the submarines as he had been instructed.

Carlson was also preparing for the withdrawal. He sent men back to the beach to uncover the rubber boats, since he was quite sure the Mille and Jaluit bases would be preparing some unpleasant surprises for him if he stayed in his current position. It would soon grow dark, and he did not want to wait for the next day.

Carlson had to have a sense of failure, and he did. He had not accomplished anything like the results he had wanted. He had hoped to overrun the island in the morning and spend the day picking up useful military information from prisoners and documents at his leisure. Instead he was engaged in a constant firefight and his men were more or less pinned down, but out at sea, four miles away, the submarines were waiting for darkness so they could come in closer for the rendezvous. It was time for the Raiders to move out.

Hoping against hope, Carlson continued to hang on a little longer until 6 PM came and the shadows were growing long.

By 6:30 it was growing dark, so reluctantly he gave the order and it was passed up front and all along the line. The Raiders were to move back to the beach, board their boats, and head out to the submarines. They had only an hour until the tide would be right for the withdrawal, and then only about two hours in which to make it.

It was completely dark by 7 PM as the rubber boats were made ready for the returning Raiders. They would have only about an hour and a half to get through the surf and rendezvous with the submarines. *Nautilus* and *Argonaut* were already moving out of the safety of deep water and into the shallows to pick up the Marines as the battle line moved back toward the beach. As each platoon arrived at the command post, led by native policemen, the officers began to count noses. Eleven men had been killed in the day's fighting, and twenty had been wounded. Carlson conferred with the chief of the native police force, Joe Miller, who had been helping the Raiders all day long. Miller promised the police would find the bodies of the dead Marines and give them an honorable burial.

Then it was time to go. The boats were taken to the edge of the surf and the men jumped in and began to paddle. The wounded went out in the first boats - PFC Le Francois, with five bullets in his shoulder, Lieutenant Charles Lamb, who had been wounded twice, and eighteen others. Lieutenant Colonel Carlson waited onshore. His boat would be the last to leave. It was part of his responsibility as leader of the battalion, although he could have left it to Major Roosevelt. Not Evans Carlson. With him remained a squad of men, the rear guard, to make sure the Japanese did not stage another banzai charge when the boats were moving and the force was at its most vulnerable.

Now the trouble really began. Carlson had known that the outboard motors' real value was going to be getting the boats out through the surf on the return trip to the submarines, but now they did not have any motors and the men would have to bring the awkward boats out by sheer strength. It was one thing to paddle a slim dugout canoe through island surf, and quite another to maneuver a floppy rubber boat. The boats headed into the towering waves. They no sooner recovered from one than another was pounding down on them, filling the craft with water. By the second wave the boats were full of water to the gunwales, and the men were alternately paddling and bailing as fast as they could.

It was nearly impossible, and only seven boats made it through the surf - five to *Nautilus*, and two more to *Argonaut*. Seventy-three of the two hundred Marines were safe, but what of the others?

The situation grew more desperate by the minute. One boat after another capsized, with the men struggling back through the surf to the island. Le Francois, the man with five bullet wounds, had lost a lot of blood, and when his boat capsized he disappeared into the heavy surf. Lieutenant Lamb also went under, but someone saw him and pulled him back to the island where he was revived. Several other Marines nearly drowned. Most had lost their weapons and equipment. After an hour the ordeal was over, and as the exhausted Marines lay on the beach they wondered what was going to happen next. Fortunately most of the wounded had made it to the submarines, but four stretcher cases and several walking wounded were still on the island. As the submarines waited, doctors and corpsmen tended to the wounded. They were scheduled to leave under cover of darkness for Pearl Harbor, but would not desert the 120 men still ashore.

Lieutenant Colonel Carlson organized his puny defenses. The remaining weapons were rounded up, and a defensive perimeter was established to prevent the Japanese from walking in on them. The perimeter had just been drawn when a Japanese patrol approached. A Marine named Jessie Hawkins issued a challenge and a firefight followed, with Hawkins killing three Japanese before getting wounded. Another Marine then dragged Hawkins through the sand to the relative safety of the beach, where his wounds were treated.

Since all of the Marines' radios had gone into the sea, communication had to be established by flashlight and Morse code. The message from the submarines was encouraging. "We will stay with you until you get off. Try to make it before Christmas."

That night the Raiders shivered on the sand and waited. Further down the beach Lieutenant Peatross and his men got into a fight with a Japanese patrol, and afterward they made it back to a submarine in their boat, but the main body was still ashore at daylight. As dawn broke, Carlson dispatched patrols to look for food and ammunition. They were also looking for Japanese, but the enemy was not eager for a fight just then. The Raiders saw only three Japanese soldiers, and they did not fire on the Marines. Later that morning Major Roosevelt took the strongest men in four boats and tried to make it through the surf. After a struggle, they succeeded. Once back aboard the submarine Roosevelt conferred with the commander. Five Raiders volunteered to return to the island in order to bring food, ammunition and weapons to the beleaguered men. They also took tow lines with them so the other boats could be towed through the surf, and had made it about halfway when suddenly Japanese scout seaplanes appeared overhead to strafe the submarines. Both submerged before they could attack. The planes then turned their attention to the five Marines in the rubber boat. They came in low on a strafing run and sank the boat. The Marines began to swim, but the Japanese came back and strafed them one by one until all five were dead.

At this point, around midmorning of the second day, Carlson and about seventy men were still holding out on the island. Their situation was desperate. They were short of food and water, and did not have enough ammunition to hold out against a sustained Japanese attack.

But there was no attack. The Japanese remaining on the island had holed up somewhere, perhaps frightened by the submarines and not knowing how many Marines were ashore. The Raider patrols returned and reported they had gone as far as On Chong's Wharf and had killed three Japanese soldiers. They did not see any other signs of opposition.

So Carlson calmly went on with his assigned mission. He led a patrol to the end of the island to get what intelligence he could about the area. There they found the dead Marines and turned them over to Police Chief Miller, who promised to bury them - and he did. When Carlson returned to Makin several months later Miller showed him the neat graves of eighteen Marines.

Once he had returned to the perimeter, Carlson dispatched to the submarines to tell them the Raiders would make another attempt that night to get past the surf after darkness fell. The fate of the one rescuing rubber boat earlier that day had convinced him it was too dangerous to try during daylight. No one knew when another flight of Japanese aircraft might appear.

With most of the day left to him, Carlson continued to collect intelligence. He took out another patrol, and they destroyed all the Japanese installations they could find. They burned down a radio station and smashed the equipment. They also blew up several antiaircraft guns, and set fire to a cache of gasoline.

Lieutenant Lamb, despite being wounded, took part in the day's activities. He spotted a sloop offshore on the lagoon side, and rowed out with two men to investigate. As they approached they heard a shot, followed by a grenade explosion. The boat had been manned by a single guard, and when he saw the Marines approaching he had committed suicide.

The wisdom of Carlson's decision not to attempt the surf until after darkness fell was confirmed later. Three more flights of Japanese seaplanes came over the island that day and attacked, but they were still operating on the original intelligence from the first planes to arrive and once again attacked the positions on the old firing line where the Marines had been on the first day. They did not find the command post.

The miracle of the day was the reappearance of Marine Le Francois with his five bullet wounds. He had been thrown out of his boat the night before and washed ashore half dead, but had survived and managed to find his way back to the command post.

Darkness came suddenly that evening, as it does in the tropics. At 7:30 the signalman was again working with his flashlight, and was soon in touch with the surfaced submarines. Carlson did not believe he could get his weakened and wounded men off the island through the surf. The alternative was to move across the island to the lagoon side, where there was none - but that would mean a dangerous voyage for the submarines into shallow water. If Japanese planes attacked, the subs would not be able to crash-dive. Plus Carlson's men were in no condition to drag their boats across the island, but they could paddle to the entrance of the lagoon if they could get through the surf and that would be a safer place to meet. So Carlson sent a Morse code message to the submarines by flashlight suggesting the lagoon rendezvous. Finally the answer came, in the form of a question.

"Who followed my father?"

That was a question no one but Evans Carlson could answer. The submarine skippers were afraid of a trap, and wanted to find out if they were in touch with Raiders or English-speaking Japanese. During the voyage from Pearl Harbor, Carlson and Navy Commander Haines had talked about many things. Since Haines' father had been a Marine officer, they had much in common. The men had argued about Marine matters, one of them being the name of the officer who had followed Haines' father as adjutant of the Marine Corps. Haines had said it was Squeegie Long. Carlson understood the question, and knew this was not the time to argue about the answer.

"Squeegie" went back the message from shore, and the fears of the submariners were allayed. They would now come around the island and brave the shallows to get the Raiders off.

The day had been hot and dry, and the sea almost calm, so the outboard motors on two of the boats were started by an optimist. Surprisingly, they ran. The boats were lashed together, and the stretcher cases loaded aboard first. Then came the walking wounded. It was nearly 9 PM before they could get going, but the able-bodied men paddled, the motors helped the boats along, and they reached the rendezvous point at about 11:30. There were no unpleasant surprises, no lights showed on the shore, and by midnight all of the men were aboard the submarines.

Once again Carlson counted noses, and this time he was in for a shock. A dozen men were missing. Carlson had thought those men were with Peatross, and Peatross had thought they were with Carlson. In the confusion of the failed attempts to reach the submarines they had been left ashore, but it was too late to go back. The submarines had to get to deep water and be prepared to dive at dawn. No one knew where the twelve men were, and there was no time to search for them.

On the way back to Pearl Harbor the two doctors and the corpsmen aboard the submarines treated the wounded. The exhausted Raiders slept their way home, while Carlson spent his hours in the wardroom considering the results of his raid. He really did not know much about it. They had sunk a 3,500-ton island trader and a patrol boat. They had burned the Japanese stocks of gasoline and wrecked their defense installations. They had destroyed two seaplanes, and killed between fifty and a hundred Japanese. There was no real way of telling. The Raiders had lost eighteen men killed, seven drowned in the surf, and fourteen wounded - plus the dozen missing men presumed to be still on the island.

When the submarines arrived back at Pearl Harbor they were greeted by a brass band and Admiral Nimitz, who had brought half his staff down to congratulate them. Their operation had been carried out in secrecy, but now that it was finished the veil was lifted and reporters were given access to the Raiders. The Navy needed some good publicity, and the American people needed a success story - for just then the battle for Guadalcanal was going very badly.

Besides publicity there were medals. Sergeant Thomason was posthumously awarded the Medal of Honor for his leadership on the battle line. Lieutenant Colonel Carlson and Major Roosevelt were each awarded the Navy Cross for their execution of the operation. And there were others.

But the best part of the story was not told, and could not be. Nimitz and his radio intelligence officers knew the raid had been more successful than anyone realized, for when the Japanese high command learned of the attack on the Gilberts, coming as it did shortly after the landings on Guadalcanal, they did not know what to make of it. They had originally thought the Guadalcanal landings were no more than a raid, and had sent only a battalion of troops to face more than a division of Marines. That move had been disastrous. So when word of the Makin landing came, the Imperial High Command diverted a task force of cruisers, destroyers and transport ships that were originally intended for the reinforcement of Guadalcanal. When they arrived, of course, they discovered the Raiders were long gone. The Carlson raid had caused the task force to be redirected, and had relieved the Americans at Guadalcanal from a new threat at a time when they were in a desperate battle to hold on. Admiral Nimitz could not say a word about it, however, because his information came from intercepted Japanese messages. To admit it would alert the Japanese that the Americans had broken their naval codes. So Lieutenant Colonel Carlson and his Raiders never got full credit for the success of their mission, which helped turn the war around in the Solomons at a time when the Japanese seemed to be winning every battle.

ISLAND FORTRESS

Tarawa

Japanese Rear Admiral Keiji Shibasaki boasted that "a million men could not take Tarawa in a hundred years." The Second Marine Division did it in just three days.

The American capture of the island of Betio in the Tarawa Atoll proved to be a series of mistakes and mishaps. Victory came because the enemy was overwhelmed by Allied superiority in men, weapons and supplies. It also came because of the heroism and tenacity of the Marines who waded ashore through the lagoon in the face of withering fire and almost certain death. Even so, Japanese forces stubbornly gave ground and in general refused to surrender.

113

Although costly, American mistakes were not forgotten and the battle upon the small island fortress would provide American planners with priceless information on the technique of island fighting.

Japan seized Tarawa Atoll at the onset of their Pacific campaign, and in December of 1941 the Japanese attacked the Pacific Fleet at Pearl Harbor and the United States entered the war. Japanese aggression shortly led to the capture and control of a large sphere of influence in the Pacific. They invaded and seized the British controlled Gilbert Islands, including Tarawa Atoll and the island of Betio. Their capture represented the Japanese southeastern most expansion in the South Pacific, and the islands served as outposts at the eastern approaches to the Japanese controlled Marshall Islands. The islands also posed a threat to Allied communications between the Central and South Pacific.

In response, Admiral Chester Nimitz assembled a fleet of land, air and sea forces under Vice Admiral Raymond A. Spruance in October of 1943 to invade and conquer the Japanese held atolls of Tarawa and Makin. Nimitz believed the capture of the Gilberts essential to any later invasion of the Marshall Islands, because without them an attack on the Marshalls would leave an invasion force susceptible to attack from the rear.

His fleet was composed of over two hundred ships and included seventeen carriers, twelve battleships, eight heavy and four light cruisers, and sixty-six destroyers. The carriers held over nine hundred planes. For the land invasion, three dozen transport ships carried the 2nd Marine Division and elements of the 27th Infantry Division. The total land invasion force numbered thirty-five thousand Marines and soldiers.

Nimitz divided the armada into three task forces - the northern task force to assault the Makin Atoll, a southern force to capture the Tarawa Atoll, and the fast-carrier force to protect the other two forces. The primary objective within the Tarawa Atoll was the island of Betio, which was two miles long and a half mile wide at its widest point. The Japanese Commanding Officer, Rear Admiral Keiji Shibasaki, had located his headquarters there and built an airstrip on the island.

The Japanese force numbered over 4,800, including 2,600 soldiers, 1,000 Japanese construction workers and 1,200 civilian laborers. The heavily fortified island was comprised of a network of steel and concrete pillboxes, tunnels, and sea walls with fourteen coastal defense guns and over one hundred machine guns. The defenders were deeply entrenched and able to withstand American naval and air bombardment.

D-Day was at 8:30 AM on November 20th, 1943. Following naval and air bombardment, Marines from the 2nd Marine Division landed on Betio. The assault took place on Red Beaches One, Two and Three, which were located on the lagoon side of the island.

The plan was for the naval and air bombardment to continue right up until the Marines hit the shore in order to take out key Japanese defensive positions, keep the Japanese forces defending the island pinned down, and provide cover for the advancing Marines. The initial assault proved to be a disaster, however, and the cover the Marines had counted on was lost due to a series of blunders by American forces. The naval and air forces failed to coordinate their schedules, and the carrier planes arrived a half hour later than expected. The amtracs and Higgins boats also got started late, so the naval bombardment had stopped long before the landing craft approached the shore. This permitted Japanese defenders to emerge from their network of tunnels and fire on the approaching Marines. The failure of the Navy to adequately research the depth of the lagoon and coral reefs along it led to further disaster for the landing craft. Higgins boats became stuck on the coral reef below, and became easy targets for the defenders. Marines inside them had to wade ashore unprotected and under constant machine gun and mortar fire. Battalions in the first waves suffered casualties as high as seventy-five percent.

At noon the situation was critical. Seeking reinforcements, General Julian Smith, who was commanding the Marine invasion onto Betio, radioed General Holland Smith, "The issue is in doubt." The general then ordered the 6th Marines' combat team dispatched from Makin into the battle at Betio.

With reserves on the way and his Marines locked in a vicious battle barely a few yards onto the shore, General Julian Smith then ordered his last reserve battalion, Major Hay's 1st Battalion, into landing craft in preparation for reinforcement of the five battalions on the island - although they would not be ordered into the battle until the next morning. This meant the depleted units on the beachhead would have to endure possible Japanese nighttime counterattacks without the help of the reserve force.

By nightfall Marine positions consisted of men along the pier and sea wall just a few yards from the water. Communications were down, and messages had to be relayed by runner. With their backs to the sea, the Marines dug in and prepared for a Japanese counterattack. Fortunately for the Marines, General Shibasaki's communication lines had been demolished during the day's initial bombardment. Without communications, the Japanese leader was unable to coordinate his troops for a successful counterattack against the Marine beachhead. Shibasaki had lost the opportunity to recapture the American position and drive the Marines back into the sea, and he wouldn't get another.

The next morning Major Hay's 1st Battalion was ordered to land, and his troops were immediately fired upon by Japanese mortars, as well as machine guns firing from an unexpected direction. During the night, several suicide-minded Japanese soldiers had swum out to the wrecked American landing craft and set up machine guns. Troops on shore were now being fired upon from the rear, and Major Hay's approaching battalion was forced to endure the assault without the benefit of strong cover. Another failure of American planners to study tide charts caused more Higgins boats to lock up on the coral reefs during high tide. Hay's rifle companies took over forty percent casualties. Allied fighter aircraft tried strafing enemy positions, but it proved futile since the enemy was heavily entrenched in pillboxes and bunkers.

As the tide rolled in, so did the Higgins boats filled with badly needed supplies. While the Marines were still pinned down on the beaches, supplies of ammunition, medical supplies and communication equipment raised the beleaguered troops' morale. With communications restored and Major Hay's troops bolstering the American beachhead, the Marines prepared for another assault and breakout. The Marines on Red Beach Two pushed their way across the airstrip and seized an inland foothold. At the same time, Major Ryan's forces on Red Beach One advanced toward Betio's southwest corner and cleared the area known as Green Beach. For the first time, a Marine beachhead had been secured. This would prove vital to the assault, as more men and supplies could come in unmolested. The Marines were no longer just maintaining a foothold on the beach - now they were able to advance.

Having just arrived from Makin, the 6th Marines were ordered to land one battalion on Green Beach at dusk. Major William Jones' 1st Battalion arrived onshore, fresh and intact. Marine assault leader Colonel David M. Shoup radioed back, "We are winning." Tanks, artillery, supplies and men now arrived on Green Beach in increasing numbers. Without any relief or fresh troops from the Imperial Japanese Fleet, Admiral Shibasaki's forces could not hold out indefinitely. His troops, however, fought on defiantly.

On the morning of D-Day-plus-2 American forces assaulted Shibasaki's headquarters. With tank support, Major Crowe's mortars finally cracked open the Japanese pillboxes protecting the command post. Shibasaki's bunkers and the Japanese soldiers within them were exterminated with brutality typical of the Pacific Campaign. Grenades dropped into the bunker's air vents caused scores of Japanese soldiers to scramble outside, only to be cut down by canister shot and rifle fire. A Marine bulldozer then covered the bunker in sand, entombing the inhabitants. Gasoline was then poured into the vents and TNT charges dropped inside. Over two hundred charred bodies were eventually discovered when the Marines entered the bunker.

In the afternoon American forces linked up on the airstrip. The arrival of 3rd Battalion, 6th Marines on Green Beach permitted Major Jones' fresh 1st Battalion to push along the southern shore of Betio and eventually link up with Marines who had earlier crossed the airfield. By nightfall the Japanese airfield and the entire western end of Betio had fallen to American forces. The remaining Japanese stronghold lay along a narrow strip of land on the eastern end of Betio.

That night the Japanese launched their last counterattack. Rather than wait to be overrun, the remaining Japanese forces made a suicidal charge against Major Jones' forces, and the line held in deadly hand-to-hand combat. The last remaining Japanese stronghold on the island had been extinguished.

On D-Day-plus-3 Betio was declared secured, and the first American aircraft landed on the island's airstrip - but the American victory at Tarawa had proved costly. Over one thousand Marines had been killed, and 2,300 were wounded. The Japanese had fought stubbornly, preferring to die rather than surrender. Less than 150 Japanese soldiers and Korean laborers surrendered. The remaining 4,700 died while fighting, or committed suicide rather than be captured. On the home front, criticism of "the Tarawa fiasco" called into question the armed forces' Pacific Campaign. Amtracs, which suffered over eighty percent casualties on Betio, were soon strengthened with better armor and weaponry. Increased and refined aerial intelligence provided accurate data concerning tidal conditions and water depths for future amphibious invasions. Deeply entrenched enemies behind bunkers were now understood to be able to withstand mass aerial and naval bombardment. Precision bombing would be needed to destroy heavily fortified positions. In the end, the lessons learned at Tarawa helped save many lives in future island battles.

The Japanese had boasted that "a million men could not take Tarawa in a hundred years. The Second Marine Division had done it in just three days.

THE HAWK

First Lieutenant William D. Hawkins

"Casualties - many. Percentage of dead - not known. Combat efficiency - we are winning."- Colonel David M. Shoup, Tarawa, 21 November 1943

Lieutenant William Deane Hawkins had a certain cowboy look about him, and was handsome despite the burn scars on his face. Everyone called him 'Hawk.' He had worked as a ranch hand and railroad brakeman before joining the Marines in 1941. The scars he had gotten as a child, from running into his mother while she was carrying a pan of hot water. Hawkins had left his home in El Paso for the last time with a distinct premonition that he would die in battle. Bidding farewell to his boyhood pal, Ballard McClesky, he had said, "Mac, I'll see you someday - but not on this earth."

He had made something of a reputation for himself on Guadalcanal, leading long-range reconnaissance patrols that stayed in the jungle for days. A private first class when he went overseas, he was quickly promoted to corporal and then sergeant, and in November of 1942 he was commissioned a second lieutenant while taking part in the Guadalcanal campaign. Then, in June of 1943, he was promoted to first lieutenant. For *Operation Galvanic* (the invasion of Tarawa), he commanded the Scout-Sniper Platoon, a select group of stalwarts who would be the first ashore on D-day. Scheduled to cross the line of departure fifteen minutes ahead of the first wave, they were to land on the end of a long pier and wipe out any machinegun positions along its seven hundred yard length. This was an important task on which the safety of the landing force depended, since any Japanese gunners emplaced on the pier would be able to enfilade the waves in both directions - while in friendly hands it could provide direct communication with the shore.

One the morning of the Tarawa landing a formation of eighty-seven amtracs churned toward the shore in three waves. Several hundred yards ahead of them were a pair of landing craft bearing Lieutenant Hawkins and his Scout-Snipers. When the first boat reached the seaplane ramp at the pier's lagoon end two Marines jumped off, followed shortly by four others, and ran onto the pier to find cover amid a stack of boxes.

The rest of Hawkins' platoon remained huddled in the two boats, awaiting his signal to follow. It never came. Hawkins found that he and his five Marines needed no help in annihilating the clusters of Japanese who were even now firing into the oncoming wave of amtracs. The nearest machinegun crew was firing from inside the equipment shack near the pier's end. Lieutenant Alan Leslie, carrying a flame thrower, aimed the nozzle and squeezed out a billowing blast of orange flame. The shack and its inhabitants vanished in a fierce crackling blaze, and the Marines rushed on past. Farther down the pier, Hawkins pointed to a half-sunk motorboat resting against the pilings on the east side. Leslie loosed another rolling ball of fire, its edges sizzling as it stroked the surface of the water, and the six moved past what was now a blackened pile of burning lumber. Thirty yards farther on Hawkins motioned his men flat, and lobbed a grenade at a machinegun crew he had glimpsed on a platform among the trestles. The explosion jogged the entire end of the wooded pier. In this fashion Hawkins, Leslie and the others cleared the pier of all snipers and machine gunners, and raced back to the Higgins boats. His mission completed, Hawkins ordered the coxswains to steer both boats out into the lagoon, where the Scout-Snipers were to wait for an emptied amtrac to carry them ashore.

Once he and his men had reached the shore, Lieutenant Hawkins led his Scout-Snipers in a series of assaults that ran parallel to the beach and forced his way into the forbidden zone near the cove later characterized by journalists as "the Pocket." The sense of urgency at this point was tremendous. It wasn't long before Lieutenant William Deane Hawkins had killed more Japanese than anyone on Betio, but he had also taken more chances.

"Aw, they can't hit me," someone heard him say. "They couldn't hit me with a shotgun at point-blank range." His disdain was unwarranted, however. He had already been wounded once.

At around two o'clock in the afternoon Hawkins showed up at Colonel David Shoup's command post.

"They won't fight, Colonel," he said, throwing down his helmet in disgust. "The bastards won't fight."

He was referring what he perceived to be the timidity displayed by a few Marine fire teams he had observed. Shoup immediately ordered Hawkins to take his Scout-Snipers and attack a series of compounds along the beach, off in the direction of the so-called Pocket. Without acknowledging that he had heard, Hawkins scooped up his helmet and sauntered away. Shoup saw that he was heading out alone and unprotected into Japanese-held territory, and cupped his hands to yell after him. Hawkins neither stopped nor turned around, but kept on walking toward a sand-covered bunker which several squads had been unable to "reduce" after repeated tries. Miraculously, Hawkins managed to complete his mad seventy-five yard walk and flip a grenade through the aperture, ducking before the mound erupted in a geyser of sand. It was the kind of performance that brought you either a Medal of Honor or death - or both. It occurred to Colonel Shoup that his swift young hero would probably not survive the day, let alone the battle.

During the down beach attacks that followed, an incoming mortar round killed three of Hawkins' men and wounded him in the hand. He pushed the corpsman away.

"I came here to kill Japs," he said, "not to be evacuated."

Hawkins made one final charge. While attacking a fort at the base of a sandy knoll, he started tossing grenades from close range. He had thrown about half a dozen when a heavy machinegun opened up and an explosive shell hit him in the shoulder. The blood just gushed out of him.

Robert Sherrod, then Editor of *The Saturday Evening Post*, wrote the following about the Marine platoon leader:

"Hawkins had told me aboard the ship that he would put his platoon of men up against any company of soldiers on earth and guarantee to win. He was wounded by shrapnel

as he came ashore in the first wave, but the furthest thing from his mind was to be evacuated. He led his platoon into the forest of coconut palms. During a day and a half he personally cleaned out six Jap machine gun nests, sometimes standing on top of a track and firing point blank at four or five men who fired back at him from behind blockhouses. Lieutenant Hawkins was wounded a second time, but he still refused to retire. To say that his conduct was worthy of the highest traditions of the Marine Corps is like saying the Empire State Building is moderately high."

Four Marines carried the dying officer to a field hospital, where the surgeon went to work on him at once - but it was too late. That night William Hawkins lay out under the slowly wheeling Southern Cross, suspended in a morphine haze. He died before dawn, and thus ended the short career of an authentic foursquare Medal of Honor hero, a strange fierce fellow, whose greatest joy was to close with the enemy.

Adapted from the book *Line of Departure: Tarawa* by Martin Russ

CACTUS AIR FORCE

Captain Joe Foss

"The story of Joe Foss's life is a story of human endeavor so great and so accomplished that it defies exaggeration."
- Senator John McCain

Joe Foss was born in 1915 to a Norwegian-Scots family in South Dakota, where he learned hunting and marksmanship at a young age. Like millions of others, eleven-year-old Joe was inspired by Charles Lindbergh, especially after he saw Lindy at an airport near Sioux Falls. Five years later he watched a Marine squadron put on a dazzling exhibition, led by Captain Clayton Jerome, the future wartime Director of Marine Corps Aviation.

In 1934 Foss began his college education in Sioux Falls, but he had to drop out to help his mother run the family farm - although he did manage to scrape up sixty-five dollars for private flying lessons. Five years later he entered the University of South Dakota again, and supported himself by waiting on tables. In his senior year he also completed a civilian pilot training program before he graduated with a Business degree in 1940.

Upon graduation, Foss enlisted in the Marine Corps reserves as an aviation cadet. Seven months later he earned his Marine wings at Pensacola and was commissioned a second lieutenant, and for the next nine months was a 'plowback' flight instructor. He was at Pensacola when the news of Pearl Harbor broke, and since he was Officer of the Day he was placed in charge of base security. Thus he prepared to defend Pensacola from Jap invaders, riding around the perimeter on a bicycle. To his dismay he was then ordered to the aerial photographer's school and assigned to VMO-1, a photo reconnaissance squadron, but insisted he wanted fighter pilot duty even after being told "You're too ancient, Joe. You're twenty-seven years old!" After lengthy lobbying with Aircraft Carrier Training Group Foss learned all about the new F4F Wildcat, logging over 150 flight hours in June and July, and when he finished training he became executive officer of VMF-121. Three weeks later Foss was on his way to the South Pacific, where Americans were desperately trying to turn the tide of war. Upon arriving in the Pacific Theater, VMF-121 was loaded aboard the escort carrier *Copahee*.

On the morning of October 9th they were catapulted off the decks in Joe's only combat carrier mission. Landing at

Henderson Field, he was told that his fighters were now based at the 'cow pasture' and was impressed with the 'make-do' character of the "Cactus Air Force." The airfield was riddled with bomb craters and wrecked aircraft, but also featured three batteries of 90mm anti-aircraft guns and two radar stations. As exec of VMF-121 he would normally lead a flight of two four-plane divisions, whenever there were enough Wildcats to go around. He was the oldest pilot in the flight, four years older than the average age of twenty-three. The flight became known as "Foss' Flying Circus" and would rack up over sixty victories. Five of them would become aces, and two would die in the in the fight for Guadalcanal.

On October 13th VMF-121 scored its first victories when Lieutenants Freeman and Narr each got a Japanese plane. Later that same day, Foss led a dozen Wildcats to intercept thirty-two enemy bombers and fighters. In his first combat a Zero bounced Foss, but overshot, and Joe was able to fire a good burst and claim one destroyed aircraft. Instantly three more Zeros set upon him, and he barely made it back to "Fighter One," with his Wildcat dripping oil. Chastened by the experience, Foss declared "You can call me 'Swivel-Neck Joe' from now on." From the first day, Foss followed the tactics of Joe Bauer - getting in close... so close that another pilot joked that the 'exec' left powder burns on his targets. The next day while intercepting a flight of enemy bombers Joe's engine acted up and he took cover in the clouds, but suddenly a Wildcat whizzed past him tailed by a Zero. Joe cut loose and shot the Zero's wing off. It was his second victory in two days.

Mid-October was the low point for the Americans in the struggle for Guadalcanal. Japanese warships shelled the U.S. positions nightly, with special attention to the airstrips. To avoid the shelling, some fliers slept at the front lines. Around that time Foss grew to appreciate the Navy's fighter doctrine, and found that the "Thach Weave" maneuver effectively countered the Zero's superior performance because "it allowed us to point eyes and guns in every direction."

Foss was leading an interception on morning of October 18th when the Zero top cover pounced on them and downed an F4F. Joe was able to get above them and flamed the nearest, hit another, and briefly engaged a third. Gaining an angle, he finally shot up the third plane's engine. Next he found a group of Bettys already under attack by VF-71 and executed a firing pass from above. Foss flashed through the enemy bombers and pulled up sharply, blasting one from below. Nine days at Guadalcanal, and he was an ace! Two days later Lieutenant Colonel Bauer and Foss led a flight of Wildcats on the morning intercept. In the dogfighting Joe downed two Zeros, but took a hit in his engine. He ended up landing safely at Henderson Field with a bad cut on his head, but was otherwise unharmed.

'Cactus Fighter Command' struggled to keep enough Wildcats airworthy to meet the daily Japanese air strikes. On the 23rd it put up two flights, led by Foss and Major Davis. There were plenty of targets, and Joe soon exploded a Zero. He then went after another which tried to twist away in a looping maneuver. Foss followed and opened up while inverted at the top of his loop, caught the Zero, and flamed it. He later described it as a lucky shot. Next he spotted a Japanese pilot doing a slow roll and fired as the Zero's wings rolled through the vertical - and saw the enemy pilot blown out of the cockpit, minus a parachute. Suddenly he was all alone and two Zeros hit him, but his rugged Grumman absorbed the damage, permitting Foss to flame one of his assailants. Once again, he nursed a damaged fighter back to

Guadalcanal. So far he had destroyed eleven enemy planes, but had brought back four Wildcats too damaged to fly.

October 25th was the day the Japanese planned to occupy Henderson Field, and they sent their fighters over with orders to circle until the airstrip was theirs. It didn't work out that way, as the U.S. ground forces held their lines and 'Cactus' did its part. Joe Foss led six Wildcats up before 10 AM, and claimed two of the Marines' three kills on that sortie. Afterwards, he berated himself for wasting ammunition on long-range shooting. He kept learning how important it was to get close. In an afternoon mission on the 25th Foss downed three more, to become the Marine Corps' first 'ace in a day.' Joe Foss had now achieved fourteen victories in only thirteen days.

Despite rugged living conditions and the stress of daily combat flying, Joe retained his enthusiasm. He and some other fliers of VMF-121 occasionally went prowling with their rifles in the jungle looking for Japanese soldiers, but Colonel Bauer stopped this activity - trained fighter pilots were too valuable to risk this way. They slept in six-man tents and ate the wretched powdered eggs that are mentioned in almost every pilot's memoirs. One guy had a gramophone that they played scratchy records on. They bathed in the Lunga River, and many grew beards rather than try to shave in cold water. They kept the beards neatly trimmed, not for appearances, but to ensure they didn't interfere with the close-fitting oxygen masks. 'Washing Machine Charlie' and 'Millimeter Mike' harassed the field nightly, so some pilots tried to sleep in the daytime.

On November 7th Foss led seven F4Fs up the Slot to attack some destroyers and a cruiser, which were covered by six Rufe floatplane fighters. They dispatched five of the Rufes promptly and prepared to strafe the destroyers. Joe climbed up to protect the others and got involved in a dogfight with a Pete, a two-man float biplane. He shot down the slow-flying plane, but not before its rear gunner had perforated the Wildcat's engine with 7.7mm machine gun fire. Once again Foss' aircraft started sputtering on the way home, but this time it didn't make it. As the engine died, he put it into the longest possible shallow dive in order to get as close to home as he could.

As the plane went into the water off Malaita Island, Foss struggled with his parachute harness and his seat. He went under with his plane, gulped salt water, and almost drowned before he freed himself and inflated his Mae West. Exhausted and with the tide against him, he knew that he couldn't swim to shore. Then, while trying to rest and re-gain his strength in his life raft, he spotted shark fins nearby. He sprinkled the chlorine powder supplied for that purpose in his emergency pack, and that seemed to help. As darkness approached, Joe heard some searchers looking for him. They hauled him in and brought him to Malaita's Catholic mission. There were a number of Europeans and Australians there, including two nuns who had been there for forty years and had never seen an automobile. They fed him steak and eggs and invited him stay for two weeks.

The next day a PBY Catalina piloted by Major Jack Cram rescued Foss. On his return to Guadalcanal he learned that 'Cactus' had downed fifteen Japanese planes in the previous day's air battle. His own tally stood at nineteen. On the ninth, Admiral Bull Halsey pinned the Distinguished Flying Cross on Joe Foss and two other pilots.

The Americans were bringing four transports full of infantry to Guadalcanal on November 12th, and in response the Japanese sent sixteen Betty bombers and thirty covering Zeroes after them while the American Wildcats and

Aircobras defended. Foss and his Wildcats were flying top cover CAP and dove headlong into the attackers, right down onto the deck. As Barrett Tillman described it in *Wildcat Aces of WWII:*

"Ignoring the peril, Foss hauled into within 100 yards of the nearest bomber and aimed at the starboard engine, which spouted flame. The G4M tried a water landing, caught a wingtip and tumbled into oblivion. Foss set his sight on another Betty when a Zero intervened. The F4F nosed up briefly and fired a beautifully aimed snapshot which sent the A6M spearing into the water. He then resumed the chase."

Foss caught up with the next Betty in line, made a deflection shot into its wingroot, and the bomber flamed up and then set down in the water. The massive dogfight continued until Joe ran out of fuel and ammunition. Between the fighters and the AA, the Americans destroyed almost all of the bombers and many of the Zeros. No U.S. ships were seriously damaged, but that night another naval surface battle raged in Ironbottom Sound. Warships on both sides were sunk or damaged, including the Japanese battleship *Hiei*, which Marine bombers and torpedo planes finished off on the 13th. The major Japanese effort continued on the 14th, as they brought in a seven-ship troop convoy. The American air forces cut this up as well.

Late that afternoon Colonel Bauer, tired of being stuck on the ground at Fighter Command, went up with Joe to take a look. It was his last flight, described by Joe Foss in a letter to Bauer's family. No trace of 'Indian Joe' was ever found. Later, back at Guadalcanal, Foss was diagnosed with malaria. Two great leaders of Cactus Fighter Command were gone, although Foss would return in six weeks.

He recuperated in New Caledonia and Australia. There he met some of the high-scoring Australian aces who viewed the Japanese as inferior opponents, and were a little dismissive of Foss' twenty-three victories. After a brief relapse of malaria, Joe returned to Guadalcanal on New Year's Day. Improvements had been made in his absence, notably the addition of pierced steel planking for the fighter strip. Foss returned to combat flying on the 15th when he shot down three more planes to bring his total to twenty-six.

Joe flew his last mission ten days later when his flight and four P-38s intercepted a force of over sixty Zeros and Vals. Quickly analyzing the situation, he ordered his flight to stay high, circling in a Lufbery. This made his small flight look like a decoy to the Japanese. Soon Cactus scrambled more fighters, and the Japanese fled. It was ironic that in one of Joe Foss' most satisfying missions, he didn't fire a shot.

A few months later he went to Washington D.C. to be decorated and begin "the dancing bear act," for the twenty-six aerial victories which equaled Eddie Rickenbacker's World War I record. He gave pep talks, made factory tours, and went on the inevitable War Bond tours. In May of 1943 President Roosevelt presented him with the Medal of Honor for outstanding heroism above and beyond the call of duty as chronicled in his Medal of Honor Citation:

"For outstanding heroism and courage above and beyond the call of duty as executive officer of Marine Fighting Squadron 121, 1st Marine Aircraft Wing, at Guadalcanal. Engaging in almost daily combat with the enemy from 9 October to 19 November 1942, Captain Foss personally shot down 23 Japanese planes and damaged others so severely that their destruction was extremely probable. In addition, during this period, he successfully led a large number of escort missions, skillfully covering

reconnaissance, bombing, and photographic planes as well as surface craft. On 15 January 1943, he added 3 more enemy planes to his already brilliant successes for a record of aerial combat achievement unsurpassed in this war. Boldly searching out an approaching enemy force on 25 January, Captain Foss led his 8 F-4F Marine planes and 4 Army P-38s into action and, undaunted by tremendously superior numbers, intercepted and struck with such force that 4 Japanese fighters were shot down and the bombers were turned back without releasing a single bomb. His remarkable flying skill, inspiring leadership, and indomitable fighting spirit were distinctive factors in the defense of strategic American positions on Guadalcanal."

Back to active duty, he served as a training advisor at the Santa Barbara Marine Corps Air Station, and then became commander of VMF-115 in the South Pacific where he met his boyhood idol, Charles Lindbergh.

After the war Foss was commissioned in the South Dakota Air National Guard, which he helped to organize. Joe then turned to politics, and was elected to the South Dakota House of Representatives. During the Korean War he returned to active duty as an Air Force Colonel, and later became chief of staff of the South Dakota Air National Guard with the rank of Brigadier General. In 1954, Foss was overwhelmingly elected Governor of South Dakota, and two years later was elected to a second term. After that he was elected the first commissioner of the American Football League, and served in that capacity until 1966. He was also president of the National Rifle Association from 1988 to 1990, and was featured in Tom Brokaw's best-seller *The Greatest Generation*. Joe Foss was much more than "just" a fighter pilot!

Visit the Joe Foss Institute on the internet at www.thefossinstitute.org

THE 'CANAL

Guadalcanal

"Where are the famous United States Marines hiding? The Marines are supposed to be the finest soldiers in the world, but no one has seen them yet?" - Japanese radio propagandist as the 1st Marine Division was steaming toward Guadalcanal (they would soon find out the answer to their question)

Shortly after defeating the Japanese at the battle of Midway, the United States decided to push into the strategically important area of the southwest Pacific. Now that Hawaii was deemed secure from immediate attack, it was time to take the fight to the Japanese. Both American commanders in the Pacific, General Douglas MacArthur and Admiral Chester Nimitz, were offensive minded, aggressive leaders who welcomed the directive that came from the Joint Chiefs on July 2nd, 1942. This directive called for parallel attacks on Rabaul Island, New Guinea, and the Solomon Islands chain. Plans were started for attacking in these areas immediately.

The plans had to be looked at from a different angle, when air recon showed that the Japanese were moving troops from Tulagi to Guadalcanal and building an airfield on the latter. These islands were next to each other in the lower Solomon chain. The Americans had been warned earlier by Australian coast watchers that the Japanese were starting to occupy Guadalcanal, an island ninety miles by twenty-five miles and covered mostly by rain forests, mountains and swamps. A Japanese airfield here would jeopardize all U.S. forces in the area. Guadalcanal had to be taken, and taken right away. Normally the island would fall under the command of MacArthur but for now the boundary between the two commands was moved, giving command of the operation to Nimitz.

Preparation and training started at a feverish pace. Nimitz assigned three carrier groups (*Saratoga, Wasp, and Enterprise*) under the command of Admiral Frank Jack Fletcher to support the operation. Fletcher was in over-all command, and Admiral Turner was in command of the landing force. This force consisted of the lst Marine Division and a regiment of the 2nd Marine Division, and was commanded.by General Archer A. Vandergrift. Lastly the operation was backed up by a joint force of American and Australian cruisers and destroyers.

At 0900 hours on August 7th, 1942, eight months to the day after the sneak attack on Pearl harbor, eleven thousand Marines landed on Guadalcanal after a lengthy naval and air bombardment. The landing was not contested by the

Japanese, and the airfield was secured that first day. Tulagi was also hit by a force of one thousand Marines, but it was a different story. The Japanese resisted fiercely, and in two days of fighting the Marines killed just about all of them.

On the second day things became more difficult for the Marines on Guadalcanal. Fletcher withdrew the carrier groups for fear of air attacks from Rabaul, and Turner did the same with the transports. The Marines were now on their own in enemy territory. To make matters worse for them, Turner's transports held much needed supplies and equipment. In addition to the supplies, there were also one thousand Marines still on the transports who would be sorely needed in the coming hours. The only naval force in the area was the patrolling ships of Task Force 44. Vandergrift put his Marines in a five mile long defensive perimeter and started to finish building the airfield with the equipment he had, plus that which the Japanese had left behind.

The Japanese Commander in the area, Admiral Mikawa, sent a naval force from Rabaul down between the islands of the Solomon Chain via a channel known as "The Slot" on the night of the 8th and hit TF 44 by surprise. In two quick battles off Savo Island the allied force lost *Canberra, Quincy, Astoria, Vincennes*, and *Chicago* with a great loss of life.

On the night of August 20th the Japanese troops who had landed earlier hit the Marine line at the Tenaru River in a fanatical "Banzai" attack. The young Marines held their ground and slaughtered the attackers, and when the sun came up the ground in front of the Marine line was littered with over eight hundred dead Japanese. These young Americans who had been civilians a short time ago had stood up to a professional, experienced army and beaten them. Their hardships and heroism were just starting though.

August 20th was also the day the first Marine fighter planes landed on the now usable airfield. They quickly dubbed themselves the "Cactus Air Force." The field itself was named Henderson Field in honor of Major Lofton R. Henderson, who had been killed in the Battle of Midway.

The Japanese kept underestimating the strength of the Americans on Guadalcanal, and continued putting their troops ashore piecemeal. They also kept up the pressure on the U.S. Navy when it returned to the area. In ensuing sea battles *Enterprise* was crippled by bombs, while the Japanese lost a seaplane carrier and over seventy planes. A Jap troop ship was also sunk, and *USS Saratoga* was put out of action for three months by torpedoes. *Wasp* and the battleship *North Carolina* were also sunk. The loss of life inflicted upon the sailors engaged in those actions was extremely high.

The battle for the island continued with the Americans landing troops and supplies during daylight hours and the Japanese doing the same after dark. This procedure, with the Japanese using ships (mostly destroyers) to shuttle troops in at night, became known to the Marines as "the Tokyo Express." On the night of the 21st of August came another Banzai attack against Henderson Field. One thousand Japanese ran screaming into the Marine positions, and eight hundred were killed before morning.

The "Tokyo Express" dropped off another six thousand troops, and on the 13th of September 3,500 of them hit the south perimeter of the airfield. This area was defended by the lst Marine Raider Battalion under the command of Lieutenant Colonel Merritt "Red Mike" Edson. They were dug in on a ridge, and bore the brunt of wave after wave of Banzai attacks. Edson was all over the field of battle, exhorting his men, and fighting right on the line with them.

At one point the Japanese breached his line and he ordered a pullback, calling in artillery strikes on their previous positions and catching the attackers in the open. This area became known as "Bloody Ridge."

Dawn broke over the bodies of one thousand dead Japanese lying in and around the Marine positions. The balance had fled back into the jungle. After the battle Vandergrift sent large patrols into the jungle after the retreating enemy, and there was almost a serious setback when a battalion of Marines were hit by a large body of Japanese and were pushed back to the beach. It looked like they'd be overrun until a destroyer responded and bombarded the attacking Japanese while the Marines were evacuated by landing craft. It was during this operation that Coastguardsman Douglas Munro put himself in harm's way while evacuating the Marines and received the Medal of Honor posthumously. He was the only member of the Coast Guard to receive this honor.

The tide began to turn against the Japanese when the "Cactus Air Force" started to operate. Now the Japanese no longer had control of the air, and soon the skies would be clear of them altogether. That was partly due to men like Marine Captain John Smith, who became the first ace of the squadron and won a Medal of Honor.

On the 18th of September the 7th Marines landed with another 4,200 men, and Vandergrift became even more aggressive. Firefights were a daily occurrence now. The Japanese were still determined to kick the Americans off the island, and were landing about a thousand men a night. The Marines kept on shooting them. The Japanese finally landed a full division on Guadalcanal under the command of General Masao Maruyama, who planned to hit the Americans in full force and put an end to them once and for all. He had his division split into two attacking forces. While one hit the Marines from the west, the other would strike from the south. The latter force would hit the Marines on Bloody Ridge again. This battle would feature two of the Marine Corps' legendary figures, Gunnery Sergeant John Basilone and Lieutenant Colonel Lewis "Chesty" Puller, and the young Marines serving with them would soon become war hardened veterans.

The all-out attack the Marines expected hit them on the night of October 24th. The brunt of the assault came against the south perimeter of Bloody Ridge again, in wave after wave of Banzai attacks. This position was held by Puller's 1st Battalion, 7th Marines. At almost 10 PM the Japanese came screaming out of the jungle and into heavy machine gun fire. Gunnery Sergeant John Basilone, set up in the middle of the line, fired a constant stream of bullets from one gun and kept the other guns supplied with ammo. He moved about the positions directing fire and had to run to the rear on several occasions to bring up more ammo. Several times he had to have his men crawl out in front of their position and drag the bodies of the dead Japanese away, because they would pile up so high as to block the field of fire. The attacks continued all night as did the rain, and when it ended there were thirteen hundred Japanese lying dead in front of the Marines, a large percentage of them killed by Basilone's machine gunners.

In November the 182nd U.S. Army Infantry Regiment was landed to bolster the Marines. The "Tokyo Express" was still landing troops each night, and the outcome was still to be decided. The war at sea was just as savage as that on land. In fact, more Americans would be killed in sea battles in this campaign than would die ashore. Shortly after midnight on November 13th a fierce surface battle erupted north of

Guadalcanal. It was one of the largest sea battles of the war. The U.S. Navy took another beating, losing *Juneau, Atlanta,* and four destroyers. The *San Francisco* was also badly damaged.

The Japanese also suffered losses. The battleship *Hei* was sunk. They also lost some twelve thousand men from their 38th Division, drowned when the U.S. sank their troop transports on the 14th of November. Navy Lieutenant Commanders Bruce McCandless and Herbert Schonland received the Medal of Honor for their actions in this sea battle. Another member of the "Cactus Air Force," Lieutenant Colonel Harold Bauer, who had received the MOH for actions on October 16th, was missing in this action and later declared KIA. Another Medal of Honor winner was Marine Captain Joe Foss, who between October and January shot down twenty-six Japanese planes.

The Tokyo Express finally petered out and came to a halt on November 30th. The 1st Marine Division was officially relieved, and the Army took over on December 9th, 1942. These men, relatively new to military service and led mostly by officers who - except for the higher grades - were also new to the military, had fought face to face with a battle hardened, experienced enemy and beaten them.

The battle now was continued by the XIV Corps, which consisted of the 2nd Marine Division and the Army's 25th and Americal Divisions under the command of Army General Patch. The fighting was still vicious, but while American strength on and around the island was building, the Japanese strength was on the wane. Attrition was wearing them down. Due to the American buildup of ships and planes, the Japanese could only supply the island with men and supplies by submarine. On January 3rd, 1943 Japanese headquarters conceded defeat and ordered the evacuation of their remaining troops from Guadalcanal, and on the 7th the last of the defeated Japanese left the island via destroyers. They left twenty-five thousand dead on the island, and between six and nine hundred pilots in the sea. Sixteen hundred Americans were killed on the island, and many more killed at sea. The rest of the Solomon Islands chain would take almost another year of fighting before being entirely in American hands.

This victory, coming after the battles of Coral Sea and Midway, showed the world that the United States had definitely recovered from the devastating damage done at Pearl Harbor and was on the way back.

PRIDE OF THE MARINES

Sergeant Albert A. Schmid

"That which does not kill you makes you stronger."
- Friedrich Nietchze

Born in 1920, Al Schmid grew up a cheerful, freckle-faced kid in the Philadelphia neighborhood of Burholme, Pennsylvania. After his mother died, Schmid was on his own. He worked on farms and at other odd jobs, and then in 1940 became an apprentice burner at the Dodge Steel Company in northeast Philadelphia, near the Delaware River.

Since he could not afford his own place, Schmid lived with fellow Dodge Steel worker Jim Merchant and his wife,

Ella Mae, in a row house on Tulip Street near the Tacoma-Palmyra bridge. While living with the Merchants Schmid met Ruth Hartley, a friend of the family who worked at a Sears department store in Philadelphia. In time Al fell in love with Ruth, whom he called "Babs."

On Sunday, December 7, 1941 Schmid was sprawled out on the floor of Jim Merchant's house looking at the paper and trying to get up the energy to get dressed for a date he had with Ruth that night. Then all of a sudden the radio stopped playing dance music, and a voice relayed the startling news that the Japanese had attacked Pearl Harbor. Thinking it was a joke, Schmid tuned in to another station. Pretty soon, they said the same thing. "All this time," Schmid remembered, "I was lying there like a dumb cluck, not thinking of it. Finally I called to Jim and said, 'Hey, Jim, the radio keeps saying there is a war with Japan - where the hell is Pearl Harbor?'" Then he got dressed and took Ruth ice skating. Ruth did not learn about Pearl Harbor (Schmid didn't tell her) until she came home later that evening.

For a day or so, Schmid could not see how the war affected him. Then things changed. He talked to Ruth about enlisting in the Marines, but she didn't take him seriously because he was always talking big. Then on December 9th 1941 he told her, "I'm in. I went down to the Custom House and signed up."

Schmid left Philadelphia on January 5th,, 1942. After recruit training at Parris Island, S.C., and further training at New River, N.C., he returned to Philadelphia on a short leave before heading for "destination unknown." He collected a bonus from Dodge Steel for his work during 1941, and used the money to buy an engagement ring for Ruth.

Soon afterward, Schmid boarded the troop transport *George F. Elliot* as part of the 11th Machine Gun Squad, Company H, 2nd Battalion, 1st Regiment, 1st Marine Division. On August 7th, 1942 the ten thousand men of the 1st Marine Division, under Major General Archer A. Vandegrift - the largest Marine force ever engaged in landing operations up to that time - assaulted Guadalcanal, beginning the first American offensive against the Japanese.

The Marines had expected a counterattack the moment they landed, but encountered no real opposition during their first two weeks. Then the Japanese sent a crack army regiment commanded by Colonel Kiyono Ichiki from Rabaul to retake Guadalcanal. Ichiki landed his elite troops on Guadalcanal on August 18th, then marched west toward Marine positions along the Ilu River (mismarked on the American maps as the Tenaru). Lieutenant Colonel Edwin Pollock's 2nd Battalion was waiting.

H Company's machinegun squad was there also. Schmid and two other Marines, Corporal Leroy Diamond and PFC John Rivers, manned a .30-caliber water-cooled machinegun inside a sandbag-and-log emplacement camouflaged with palm fronds and jungle greenery. The position was on the west bank of the Ilu, which was fifty yards wide at that point.

At 3 AM on August 21st Ichiki, confident of victory, attacked by the green light of flares. The Japanese yelled, jabbered and fired machineguns, trying to force the Marines to reveal their positions. The Marines held their fire.

Across the river from their nest, Schmid saw a dark, bobbing mass at the edge of the water. "It looked like a herd of cattle coming down to drink," he remembered. Fifty Japanese crossed the river yelling, "Marine, tonight you die," and "Banzai!" while firing their rifles as they came.

147

Johnny Rivers opened up on them, and the mass broke up. Screams of rage and pain came from the other side as the Japanese concentrated everything they had on Schmid's position and on another machine-gun position 150 yards downstream. Bullets whined past the Marines' heads, throwing mud and wood chips around them. Schmid's heart pounded rapidly.

The machinegun on their right stopped firing, put out of action. Then a dozen bullets tore into Rivers' face, killing him. His finger froze on the trigger, sending two hundred rounds into the darkness. Cold rage rising in him, Schmid shoved Rivers' body out of the way and took over the gun. Corporal Diamond got in position to load it for him.

Every time Schmid raked the attacking Japanese he heard them yelling as bullets ripped into them. He heard one particular Japanese officer "screeching and barking commands at the others. He had a nasty shrill voice that stood out over the others." Schmid fired a burst at the voice, but failed to silence it. It would haunt him for years.

Diamond then was hit in the arm, the bullet knocking him partially across Schmid's feet. He could not load anymore, but while Schmid fired the gun Diamond stood beside him, spotting targets. Schmid would fire across the river to the left, and then feel Diamond hitting him hard on the arm and pointing to the right. He would then swing the gun in that direction and hear Japanese yelling as his bullets hit them.

Schmid now was both loading and firing the machinegun. When he got close to the end of a 300-round belt of ammunition, Diamond would punch his arm. Schmid would fire a burst, rip open the magazine, insert a new belt and resume firing. At one point a Japanese soldier put a string of bullets through the .30 caliber's water jacket. Water spurted over Schmid's lap and chest, and the gun crackled and overheated - but did not jam.

Schmid continued loading and firing the machinegun for more than four hours, with and without help. Then somehow a Japanese soldier got through the body-choked stream and got close enough to throw a hand grenade into their position.

"There was a blinding flash and explosion," Schmid recalled. "My helmet was knocked off. Something struck me in the face." When he put his hand up, all he felt was blood and raw flesh. Then he felt pain in his left shoulder, arm and hand. He could see nothing, and collapsed on his back in the nest. "They got me in the eyes," he muttered to Diamond, who lay beside him.

The Japanese were still pouring bullets into the machinegun position, and Schmid reached around to his holster and took out his .45. Diamond heard him fussing with it and yelled, "Don't do it, Smitty, don't shoot yourself."

"Hell, don't worry about that," Schmid said. "I'm going to get the first Jap that tries to come in here!"

"But you can't see," Diamond reminded him.

"Just tell me which way he's coming from and I'll get him," Schmid replied.

Both men were helpless in the hole, and it was getting light. A sniper in a tree across the river was firing almost straight down at them. The only thing protecting the two Marines was the shelf where the machinegun stood, about two feet in diameter.

Although his sight had not come back, Schmid took his position between the spread rear tripod legs of the machine gun, squeezed the trigger and, with Diamond yelling directions in his ear, resumed firing at the Japanese.

Private Whitey Jacobs, one of the squad's members, braved the continuous Japanese gunfire, jumped into the nest and treated Schmid's and Diamond's wounds. The next thing Schmid knew, they were taking him out on a blanket. He still had the .45 automatic in his hand. Hearing his lieutenant's voice, Schmid held out the weapon. "I guess I won't need this anymore, sir," he said. Then Schmid passed out.

All night the Japanese continued their assaults, but the Marines' anti-tank guns, machineguns and artillery cut Ichiki's men down. At dawn, when it was clear the position would hold, Vandegrift sent a reserve battalion across the river to attack the Japanese from their flank and rear. Of the eight hundred Japanese who attacked across the Ilu on August 21st, only fourteen wounded were picked up, and only one of those was captured unhurt. The rest were killed. Ichiki burned his regimental colors and committed suicide. The number of bodies counted within range of Al Schmid's machinegun ran into the hundreds. The other Marines who were there that night credited him with killing at least two hundred Japanese.

Schmid was put on a hospital ship and sent back to the United States. He was admitted to the Naval Hospital at San Diego, California in October of 1942, where he endured many operations to remove shell fragments from his face and eyes. His recovery was helped by the care and understanding of Virginia Pfeiffer, a Red Cross worker in the hospital, who wrote a four-page letter to Ruth explaining Schmid's wounds. "Today he told me he might as well let you know," she wrote. "He has lost one eye, and the other is seriously damaged. The doctors will not know for several months whether he will have any sight in that eye." Virginia encouraged Ruth to keep writing to Schmid. On February 18th, 1943 Al Schmid received the Navy Cross "for extraordinary heroism and outstanding courage." He went to

Washington, D.C., and was commended by President Franklin D. Roosevelt and the Joint Chiefs of Staff.

Back in Philadelphia a parade was given in Schmid's honor, and the *Philadelphia Inquirer* presented him with its 'Hero Award' and one thousand dollars. In New Orleans, he received the key to the city. Articles about him appeared in *Life* and *Cosmopolitan* magazines, and a book, *Al Schmid - Marine*, was written by Roger Butterfield. In 1944 Warner Brothers studio began production on a movie based on Butterfield's book, *Pride of the Marines*, starring John Garfield.

Before he began the movie Garfield went to Philadelphia, met the real Al Schmid, became his friend, lived in his home, and studied him. Garfield also spent two weeks at the San Diego Naval Hospital studying the characteristics and mental attitudes of blind casualties. *Pride of the Marines* was released in 1945 and became an instant hit.

Schmid never thought of himself as a hero. "When I came back I was the most disgusted man you ever saw. I didn't want to bother to do anything. I could see people looking away from my ugly scars. They wouldn't want to associate with me. I even told my girl it was all over."

Ruth would not take no for an answer, she and Al were married in April of 1943, and in June of 1944 she gave birth to a son. The publicity generated by the marriage had brought a flood of requests for war bond, hospital and charity appearances. Although he didn't want to go, Schmid accepted all of the invitations. "I wanted to help the boys, and at the same time I was helping myself," he explained. "I got used to people again. Any time anyone wanted me, I was there, whether there was a little profit or all for charity."

Al Schmid was honorably discharged from the Marine Corps Reserve in December of 1944. He died of bone cancer on December 2nd, 1982 in St. Petersburg, Florida and was buried with full military honors in Arlington National Cemetery.

MITCH

Platoon Sergeant Mitchell Paige

"One ought never to turn one's back on a threatened danger and try to run away from it. If you do that, you will double the danger. But if you meet it promptly and without flinching, you will reduce the danger by half."
– Sir Winston Churchill

About 0200, in a silence so pervasive that men many yards apart could hear each other breathing, I began to sense movement all along the front and deep in the jungle below us and to our left. We could hear the muffled clanking of equipment and periodically, voices hissing in Japanese.

These were undoubtedly squad leaders giving their instructions. At the same time, small colored lights began flicking on and off throughout the jungle. I could hear Price whispering for me to come to his foxhole. I quietly crawled over to him and he had an excellent view of someone flicking a light on and off. Price said, "I thought I was cracking up seeing all those fireflies." I assured him he was not cracking up because those were lights handled by Japanese soldiers.

As I crawled around I told the men to glue their eyes and ears to anything that moved, and reminded them that the small lights we were seeing were assembly signals for the enemy squads. I also instructed everyone not to fire their weapons, as the muzzle flashes would give away our positions and we would be raked with fire and smothered with grenades. We had to let them get closer as we were outnumbered, but when things started popping I urged each man to just hang on. Earlier Jonjock, Swanek and I had stretched a piece of wire out in front of our position and hung several empty blackened ration cans on it. We put an empty cartridge case in each can which would rattle if hit by someone's foot.

I had previously requested an artillery and mortar concentration. This was, however, denied because the enemy was still in the jungle where the effect would almost be nil. I then returned to my foxhole. Manning my number-two gun was Corporal Raymond 'Big Stoop' Gaston and Private Samuel 'Muscles' Leiphart. Their gun was at the part of our line which bordered on the side where the jungle came up to meet the ridge. They both whispered to me that there was considerable rustling very near to them in the undergrowth. I said, "Hold your fire."

Corporal Richard 'Moose' Stanberry arranged several grenades in a neat row in front of him, and then nervously rearranged them. He was fond of his Thompson submachine gun, and I never worried about him as he was well-trained - a perfectly disciplined Marine who could handle himself in any situation. Now everyone was straining to hear and see.

The bushes rustled and the maddening voices continued their soft mutterings, but still nothing could be seen. Then I dimly sensed a dark figure lurking near Gaston's position. I grabbed a grenade, pulled the pin and held the lever ready to throw it. Around me I could hear the others also pulling pins as we had the night before. We heard the ration cans rattle and then somebody let out a shriek and instantaneously the battle erupted. Grenades were exploding all over the ridge nose. Japanese rifles and machineguns fired blindly into the night and the first wave of enemy troops swarmed into our positions from the jungle flanking Gaston's gun.

Stansberry was pulling the pins out of his grenades with his teeth and lobbing them down the slope into the jungle. Leiphart was skying them overhead like a baseball pitcher. The tension burst like a balloon and many men found themselves cursing, growling, and screaming like banshees. The Japanese were yelling "Banzai!" and "Blood for the Emperor!" Stansberry, in a spontaneous tribute to President Roosevelt's wife, shouted back, "Blood for Eleanor!"

The battleground was lit by flashes of machinegun fire, pierced by the arching red patterns of tracer bullets, shaken by the blast of shells laid down no more than thirty yards in front of the ridge by Captain Louis Ditta's 60mm Mortars. It was a confusing maelstrom, with dark shapes crawling across the ground or swirling in clumped knots, and struggling men falling on each other with bayonets, swords and violent oaths. After the first volley of American grenades exploded, the wave of Japanese crowding onto the knoll thickened. PFC Charles H. Lock was killed from a burst of enemy machine-gun fire.

I screamed, "Fire machineguns! Fire!" and with that all of the machineguns opened up along with all the rifles and tommy guns. In the flickering light, I saw a fierce struggle taking place for the number-two gun. Several Japanese soldiers were racing toward Leiphart, who was kneeling and already hit. I managed to shoot two of them while the third lowered his bayonet and lunged.

Leiphart was the smallest man in the platoon, weighing barely 125 pounds. The Japanese soldier ran him through, with the force of the thrust lifting him high in the air. I took careful aim and shot Leiphart's killer.

Gaston was flat on his back, scrambling away from a Japanese officer who was hacking at him with a two-handed Samurai sword and grunting with the exertion. He tried desperately to block the Samurai sword with a Springfield he had picked up off the ground, apparently Leiphart's. One of his legs was badly cut from the blows, and the rifle soon splintered. The Japanese officer raised his sword for the killing thrust and Gaston, with maniac strength, snaked his good leg up and caught his man under the chin with his boondocker in a violent blow that broke the Japanese's neck.

The attackers ran past Gaston's gun and spread out, concentrating their fire on the left flank gun, manned by Corporal John Grant, PFC Sam H. Scott and Willis A. Hinson. Within minutes, Scott was killed and Hinson was wounded in the head. Then Joseph A. Pawlowski was killed. Stansberry, who had been near me, was hit in the shoulder, but the last time I saw him he was still firing his tommy gun with ferocity and shouting, "Charge! Charge! Blood for Eleanor!"

Corporal Pettyjohn, who was on the right, cried out in anguish, "My gun's jammed!" I was too busy to answer his call for help. At the center, we were beating back the seemingly endless wall of Japanese coming up the gentle slope at the front of the position. There were at that point approximately seventy-five enemy soldiers crashing through the platoon, most of them on the left flank, but the main force of the attack had already begun to ebb. The ridge was crowded with fighting men.

For some reason I vividly recall putting up my left hand just as an enemy soldier lunged at me with a fixed bayonet. He must have been off balance, as the point of the bayonet hit between my little finger and the ring finger just enough to let me parry it off, and as he went by me he dropped dead on the ground.

The enemy started to melt back down the slope, and almost before they were out of sight Navy Corpsmen began moving forward to treat the wounded. At Pettyjohn's gun, James 'Knobby' McNabb and Mitchell F. 'Pat' Swanek were badly wounded and had to be moved off the line. Stansberry was still around, and didn't want to leave. I crawled over to Pettyjohn's gun.

"What's wrong with it?"

Pettyjohn said it was "a ruptured cartridge which refused to budge."

I said, "Move over," and fumbled with stiff fingers, broke a nail completely off, but somehow pried the slug out with a combination tool which I found in the spare parts kit under the tripod. I also changed the belt feed pawl, which had been damaged in the rough slamming trying to get the round out, while Pettyjohn and Faust covered me.

Though the first assault had flopped, a number of enemy soldiers had shinnied to the top of the tall hardwood trees growing up from the jungle between the platoon and Fox Company's position. From this vantage point they could direct a punishing, plunging fire down in two directions. The men in the foxholes along the crest were especially vulnerable - Bob Jonjock and John Price were wounded and helped to the back of the line by the corpsmen.

I was getting ready to feed a new belt of ammunition into Pettyjohn's gun. My left hand felt very slippery, so I rubbed it in the dirt under the tripod. Then, as I reached up to hold the belt again, I felt a sharp vibration and a jab of hot pain in my hand. I fell back momentarily and flapped my arm and stared angrily at the gun, which had been wrecked by a burst of fire from a Japanese Nambu light machine gun.

Almost immediately a second assault wave came washing over our positions. This attack was more successful than the first. Oliver Hinkley and William Dudley were wounded. Hinson, over on the left gun and already wounded, continued to fire until all his supporting rifles were silenced. He then withdrew down around the hill to the rear of George Company, putting the gun out of action before he left as I had instructed.

That section had been hit hard with mortars and grenades, causing severe shock to all the men, with one of the first being August Marquez. All the men on the spur had been literally blasted off, including Lieutenant Phillips, Bill Payne and John Grant.

In the Fox Company area toward my left rear, I saw Fox Company men pulling out and disappearing over the crest. I picked up a Springfield and fired a shot at them, yelling for them to hold the line.

The Japanese swarmed up that seventy-foot cliff in great numbers, armed with three heavy and six light machine guns, a number of tommy guns and several knee mortars. I thought, "Dear God, Major Conoley and his small command post are just over the crest," but here was the only grazing fire I had with my machinegun, so I quickly found Gaston's gun and swung it around toward our own lines as there was nothing between my gun and the crest but enemy soldiers.

I fired a full belt of ammunition into the backs of those crouching enemy, praying that they could not get over the crest to the command post. I learned later from Captain Farrell, who was with Colonel Hanneken's command post, that the word was the enemy had one of Paige's fast firing machine guns and the rounds were ricocheting over the line at Major Conoley's position. He had also heard reports that all my men had been killed, and that in fact someone had seen me sprawled out dead on the ground before they left the ridge. I learned later, too, that this information had gotten back to the Division Command Post.

By 0500 the enemy was all over the spur and it appeared they were going to roll up the entire battalion front. A second prong of the attack aimed at our front had not fared as well, but my platoon was being decimated. A hail of shrapnel killed Daniel Cashman, and Stansberry had been pulled back over the hill after being hit again.

I continued to trigger bursts until the barrel began to steam. In front of me was a large pile of dead bodies. I ran around the ridge from gun to gun trying to keep them firing, but at each emplacement I found only dead bodies. I knew then I must be all alone.

As I ran back and forth, I bumped into enemy soldiers who were seemingly dashing about aimlessly in the dark. Apparently they weren't yet aware they had almost complete possession of the knoll. As I scampered around the knoll, I fired someone's Springfield that I happened to pick up. Then, somehow, I stumbled over into the right flank of George Company. There I found a couple of men I knew named Kelly and Totman. They had a water-cooled machinegun. I told them I needed their gun. At the same time, I grabbed it and they took off with me.

I said, "Follow me!" and ordered several riflemen to fix bayonets and to follow us to form a skirmish line back across the ridge. I told the riflemen not to be afraid to use their bayonets. We still had the 1905, sixteen-inch bayonets which had the front end sharpened its full length and the back edge five inches from the point.

It was by then not quite as dark as it had been. Soon dawn would break. I knew that once the Japanese realized how much progress they had made, a third wave of attackers would come up the slope to solidify their hold on the hill.

On the way back I noticed some movement of Japanese on the ridge just above Major Conoley's position, which I had raked with grazing fire earlier. I fired Kelly and Totman's full belt of 250 rounds into that area and once again the rounds were ricocheting over Conoley's head, but he had no way of knowing that I was the one doing the firing. He could only surmise that the enemy was now using our machineguns.

As we advanced across the ridge, some of the Japanese began falling back. Several of them, however, began crawling awkwardly across the knoll with their rifles cradled in the crooks of their arms. Then I saw with horror that they were headed toward one of my guns, which was now out in the open and unmanned.

Galvanized by the threat, I ran for the gun. From the gully area, several Japanese guns spotted me and swiveled to rake me with enfilading fire. The snipers in the trees also tried to bring me down with grenades, and mortars burst all around me as I ran to that gun. One of the crawling enemy soldiers saw me coming and jumped up to race me to the prize. I got there first, and jumped into a hole behind the gun. The enemy soldier, less than twenty-five yards away, dropped to the ground and started to open up on me. I turned the gun on the enemy and immediately realized it was not loaded. I quickly scooped up a partially loaded belt lying on the ground and with fumbling fingers, started to load it.

Suddenly a very strange feeling came over me. I tried desperately to reach forward to pull the bolt handle back to load the gun, but I felt as though I was in a vise. Even so I was completely relaxed, and felt as though I was sitting peacefully in a park. I could feel a warm sensation between my chin and my Adam's apple. Then all of a sudden I fell forward over the gun, loaded it, and swung it at the enemy gunner at the precise moment he had fired his full thirty-round magazine at me and stopped firing.

For days later I thought about the mystery and somehow I knew that the 'Man Above' also knew what had happened. I never wanted to relate this experience to anyone, as I did not want to ever have anyone question it.

I found three more belts of ammunition and quickly fired them into the trees and all along the ridge. I sprayed the terrain with the remaining rounds, clearing everything in sight. All the Japanese fire in the area was apparently being aimed at me, as mine was the only automatic weapon firing from a forward position. The barrage, concentrated on the ridge nose, made me feel as if the whole Japanese Army was firing at me.

I was getting some help from our mortars with the George Company Commander, Captain L.W. Martin, observing.

These rounds were laid on the spur and prevented the enemy from moving up, and without them they would have probably enveloped me from the rear. Other than this I was alone, as my George Company friends were still some distance behind me.

In addition to being in this position, I had an immediate need of more ammunition - and couldn't see any more lying around. Just at that time, aid came which made me glow with pride. Three men of my platoon voluntarily crossed the field of fire to resupply me.

The first one came up and just as he reached me he fell with a bullet in the stomach. Another one then rushed in and was hit in the groin just as he reached me. He fell against me, knocking me away from the gun. Seconds later Bob Jonjock, who had also been wounded earlier, came from somewhere with more ammunition. Just as he jumped down beside me to help load the gun, I saw a piece of flesh fly off his neck. He had been hit by an enemy bullet.

I told him to get back while I sprayed the area. He refused to leave. I said, "Get the hell back, Jonjock!" and he again said, "No, I'm staying with you."

I hated to do it, but I punched him on the chin hard enough to bowl him over and finally convince him I wanted my order obeyed. He somehow made his way back. I was afraid he would bleed to death.

Meanwhile Major Conoley, at the forward command post, was rounding up a ragtag force with which to retake the Fox Company spur. There were bandsmen serving as stretcher bearers, wiremen, runners, cooks, even mess men who had brought some hot food up to the front lines during the night and stayed just in case. Those men, numbering just twenty-four, mounted a counterattack up over the crest line which I had fired some five hundred rounds at. They found the

Japanese machineguns and several of Fox Company's weapons, including three light machineguns, all in good working order. That counterattack found ninety-eight dead on the spur by actual count.

That was at about 0530 or so. Dawn was already breaking. I was able to observe the progress of that charge from my position, as I was directly out to their front. I also watched quite a few enemy soldiers scrambling back into the jungle, but I couldn't fire in that direction. As I watched that beautiful charge, it gave me the inspiration to get up and yell to my George Company fighters with their fixed bayonets to stand by to charge. I yelled out in Japanese to stand up, "Tate! --- tah- teh, tah-the," and then said hurry, "Isoge! --- ee-soh-geh, ee-soh-geh!"

Immediately a large group of Japanese soldiers, about thirty in all, popped up into view. One of them looked quizzically at me through field glasses. I triggered a long burst and they just peeled off like grass under a mowing machine.

At that point, I turned around to tell my friends I was going to charge over the knoll and said, "I want every one of you to be right behind me," and they were. I threw the two remaining belts of ammunition that my men had brought me over my shoulder, unclamped the heavy machinegun from the tripod and cradled it in my arms. I really didn't notice the weight, which was about a total of eighty pounds, and was unaware that the water jacket of the gun was red hot.

I fed one of the belts into the gun and started forward down the slope, scrambling to keep my feet, spraying a raking fire all about me. There were still a number of live enemy soldiers on the hillside in the tall grass, pressed against the slope. I must have taken them by surprise, as the gun cut them all down. I noticed one of them was a field grade officer who had just expended the rounds in his revolver and was reaching for his two-handed sword. He was no more than four or five feet from me when I ran into him head on.

The skirmishers followed me over the rim of the knoll, and they too were all fired up and were giving the rebel yell, shrieking and cat-calling like little boys imitating Marines, sounding like there were a thousand rather than a mere handful.

They followed me all the way across the draw with fixed bayonets to the end of the jungle, where long hours before the Japanese attacks had started. There we found nothing left to shoot at. The battle was over.

The jungle was once again so still that if it weren't for the evidence of dead bodies, the agony and torment of the previous hours, the bursting terror of the artillery and mortars rounds, and the many thousands of rounds of ammunition fired, it might only have been an awful bad dream.

It was a really strange sort of quietness. As I sat down soaked with perspiration with steam still rising from my hot gun, Captain Louis Ditta, another wonderful officer who had joined the riflemen in the skirmish line and had earlier been firing his 60mm mortars to help me, slapped me on the back as he handed me his canteen of water as he kept saying, "tremendous, tremendous!" He then looked down at his legs. We could see blood coming through his dungarees. He had a neat bullet hole in his right leg.

There were hundreds of enemy dead in the grass, on the ridge, in the draw, and in the edge of the jungle. We dragged as many as we could into the jungle, out of the sun. We buried many and even blasted some of the ridge over them to prevent the smell that only a dead body can make on a hot day. A corpsman sent by Captain Ditta smeared my whole left arm with salve of some kind. He cleaned off the bayonet gash, which had since filled with dirt, and the bullet nicks on my hands which were also filled with dirt and coagulated blood. He also stuck a patch on my back just below the shoulder blade. (In 1955, I felt something irritating in my back and ended up having a piece of metal about 3/4 of an inch long removed - right where the corpsman had placed that patch.)

As the corpsman left he said, "You know, you have some pretty neat creases in your steel helmet."

I replied, "Yes, thank God - made in America."

Excerpted from *A Marine Named Mitch*

BAA BAA BLACK SHEEP

Colonel Gregory "Pappy" Boyington

"Just name a hero, and I'll prove he's a bum."
– "Pappy" Boyington

Undoubtedly the most colorful and well known Marine Corps ace was Gregory "Pappy" Boyington, Commanding Officer of VMF-214. Stories about Boyington are legion, many founded in fact, including how he led the legendary Black Sheep squadron and served in China as a member of the American Volunteer Group - the famed Flying Tigers. He spent a year and a half as a Japanese POW, was awarded the Medal of Honor, and was recognized as the Marine Corps' top ace. Always hard-drinking and hard-living, Pappy's post-war life was as turbulent as his wartime experiences.

Born on December 4, 1912, young Greg had a rough childhood - divorced parents, alcoholic step-father (whom Greg believed to be his natural father until he entered the Marine Corps), and lots of moves. He grew up in St. Maries,

Idaho, a small logging town, and got his first ride in an airplane when he was only six years old. The famous barnstormer, Clyde Pangborn, flew his Jenny into town, and Greg wangled a ride. What a thrill for a little kid! Greg's family then moved to Tacoma, Washington in 1926. While there in high school he took up a sport that he would practice for many years - wrestling. When he'd had a few too many drinks (which was often), adult Boyington would challenge others to impromptu wrestling bouts, frequently with injurious results. He enrolled at the University of Washington in 1930, where he continued wrestling and participated in ROTC. He met his first wife Helene there, and they were married not long after his graduation in 1934. His first son, Gregory Clark Boyington, was born ten months later.

After a year with Boeing, Greg enlisted in the Marine Corps. It was then, when he had to supply them with his birth certificate, that he learned of his natural father. He then began elimination training in June of 1935, where (in the small world of Marine aviation at that time) he met Richard Mangrum and Bob Galer, both future heroes at Guadalcanal. He passed, and received orders to begin flight training at Pensacola NAS in January of 1936 with class 88-C. There he flew a floatplane version of the Consolidated NY-2. Like another great ace, Gabby Gabreski, Boyington had a tough time with flight training and had to undergo a number of rechecks.

Until he arrived in Pensacola, Boyington had never touched alcohol. But there, with hard-partying fliers and aware of his wife's "fooling around," he soon discovered his affinity for liquor. Early on, Boyington established his Marine Corps reputation: hard-drinking, brawling, well-liked, and always ready to wrestle at the drop of a hat. He kept flying all through 1936, slowly progressing toward earning his wings and flying more powerful planes like the Vought O2U and SU-1 scouting biplanes. At Pensacola he also met his future nemesis, Joe Smoak, memorialized in *Baa Baa Black Sheep* as "Colonel Lard." He finally won his coveted wings in March of 1937, becoming Naval Aviator #5160.

Before reporting for his assignment with VMF-1 at Quantico, Virginia Boyington took advantage of his thirty-day leave to return home and reconcile with his wife Helene, who became pregnant with their second child. In those days Marine aviators were required to be bachelors, so Greg's family was a secret that he kept from the brass - but he brought them with him to Virginia and installed them quietly in nearby Fredericksburg. He flew F4B-4 biplanes during 1937, taking part in routine training, an air show dubbed the "All American Air Maneuvers," and a fleet exercise in Puerto Rico.

In July of 1938 he moved to Philadelphia to attend the Marine Corps' Basic School for ten months. Apparently not motivated by the "ground-pounder" curriculum, Boyington here evidenced the weaknesses that would haunt him: excessive drinking, borrowing money and not repaying it, fighting, and poor official performance.

His irresponsibility, his debts, and his difficulties with the Corps continued to mount throughout 1939 and 1940 when he flew with VMF-2 while stationed at San Diego. One memorable, drunken night he tried to swim across San Diego Bay, and wound up naked and exhausted in the Navy's Shore Patrol office. Despite his problems on the ground, it was during these days of 1940 while flying with VMF-2 that Boyington first began to be noticed as a top-notch pilot. Whatever his other issues, he could out-dogfight almost anyone - but back at Pensacola in January of 1941 his problems mounted when he decked a superior officer in a fight over a girl, and his creditors sought official help from the Marine Corps. Greg's career was a hopeless mess by late 1941.

Rescue came from, of all places, China. Anxious to help the Chinese in their war against Japan, the U.S. government arranged to supply fighter planes and pilots to China under the cover of the Central Aircraft Manufacturing Company (CAMCO). CAMCO recruiters visited U.S. military aviation bases looking for volunteers. As Bruce Gamble described it in *Black Sheep One*:

"The pilots were volunteers only in the sense that they willingly quit their peacetime job with the military; otherwise they were handsomely paid through CAMCO. Pilots earned $600 a month and flight leaders $675, plus a fat bonus for each Japanese plane destroyed. This was double or even triple the current military salary for pilots... in March, CAMCO representatives began recruiting military pilots for what would become the American Volunteer Group (AVG)... one recruiter set up an interview room in San Diego's San Carlos Hotel, a popular watering hole for pilots, and on the night of August 4 Greg Boyington found himself in the hotel bar simply 'looking for an answer.' Payday had been just a few days earlier, but already he was broke. His wife and children were gone, he was deeply in debt, and many of his superiors were breathing down his neck."

The money looked very good to Boyington. Assured that the program had government approval and that his spot in the Corps was safe, he signed on the spot and promptly resigned from the Marine Corps. While the AVG deal for pilots normally did contemplate a return to active U.S. military service, in Greg's case his superiors took a different view. They were happy to be rid of him, and noted in his file that he should not be reappointed.

He shipped out of San Francisco on September 24, 1941 aboard the *Boschfontein* of the Dutch Java Line. After docking at Rangoon, the AVG fliers arrived at their base at Toungoo on November 13th. Boyington flew several missions during the defense of Burma, and after Burma fell returned to Kunming and flew from there until the Flying Tigers were incorporated into the USAAF. His autobiography includes many war stories from his experiences with the Flying Tigers.

Boyington clashed with the leader of the Flying Tigers, the strong-willed Claire Chennault, and quit the AVG in April of 1942. Chennault gave him a dishonorable discharge, and Greg went back to the U.S. While there he claimed to have shot down six Japanese fighters, which would have made him one of the first American aces of the war. He maintained until his death in 1988 that he did in fact have six kills, and the Marine Corps officially credits him with those victories.

While with the Flying Tigers Greg also made the acquaintance of Olga Greenlaw, the XO's beautiful wife who, in her own words, "knew how to get along with a man if I liked him." Apparently she and Boyington "got along." She even wrote a book, *The Lady and the Tigers*, in 1943.

Boyington subsequently returned to the States in the spring of 1942 and took up with a woman named Lucy Malcolmson - his first marriage having fallen apart. Then with some finagling, undoubtedly helped by the wartime demand for experienced fighter pilots, he was reappointed to the Marines in November with the rank of Major. In January of 1943 Boyington embarked on the *Lurline*, bound for New Caledonia, where he would spend a few months on the staff of Marine Air Group (MAG)-11. Here he got his first close look at a Corsair, flown by his friend Pat Weiland.

Boyington finally secured assignment to VMF-122 as Executive Officer for a combat tour, but as usual he clashed with his superior - this time Major Elmer Brackett. In any event Brackett was shortly removed and Boyington took over, but did not see much action. It was at this time, in early 1943, when as the new CO of VMF-122 his claim of six kills with the AVG first made it into print. Then Smoak relieved him of his command of VMF-122 in late May, and that was followed by a broken leg and time in the hospital.

In the summer of 1943, as Boyington convalesced, the U.S. naval air forces needed more Corsairs in the fight. Oddly, the key pieces - trained pilots and operational aircraft - were present in the South Pacific, but many of them were dispersed. Who got the idea remains unclear, but he was given the assignment to pull together an ad hoc squadron from available men and planes. Originally they formed the rear echelon of VMF-124, but eventually these twenty-six pilots would become the famous "Black Sheep." In a complex, and common, wartime shuffling of designations, Boyington's team was redesignated VMF-214, while the exhausted pilots of the original VMF-214 (nicknamed the Swashbucklers) were sent home.

Under Boyington as CO and Major Stan Bailey as Exec, they trained hard at Turtle Bay on Espritu Santo, especially the pilots who were new to the Corsair. While leading this group of young pilots, most in their early twenties, Boyington - at the "advanced" age of thirty - picked up the nickname "Gramps." The Black Sheep don't remember calling him "Pappy"- that was a nickname the press picked up after he was shot down.

In early September of 1943 the new VMF-214 moved up to their new forward base in the Russells, staging through Guadalcanal's famed Henderson Field. The "Black Sheep" fought their way to fame in just eighty-four days, piling up a record 197 planes destroyed or damaged, troop transports and supply ships sunk, and ground installations destroyed in addition to numerous other victories. They flew their first combat mission on September 14, 1943, escorting Dauntless dive bombers to Ballale, a small island west of Bougainville where the Japanese had a heavily fortified airstrip. There they encountered heavy opposition from enemy Zeros. Two days later, in a similar raid, "Pappy" claimed five kills, his best single day total. In October VMF-214 moved up from their original base in the Russells to a more advanced location at Munda. From here they were closer to the next big objective - the Jap bases on Bougainville. On one mission over Bougainville, according to Boyington's autobiography, the Japanese radioed him in English, asking him to report his position and so forth. Pappy played along, but stayed five thousand feet higher than he had told them, and when the Zeros came along the Black Sheep blew twelve of them away.

During the period from September 1943 to early January 1944 Boyington destroyed twenty-two Japanese aircraft. By late December it was clear he was closing in on Eddie Rickenbacker's record of twenty-six victories (including his six with the AVG), and the strain was starting to tell. On January 3, 1944 Boyington was shot down in a large dogfight in which he claimed three enemy aircraft, and was captured. The following is an excerpt from Boyington's book *Baa Baa Black Sheep* describing his final combat mission:

"It was before dawn on January 3, 1944, on Bougainville. I was having baked beans for breakfast at the edge of the airstrip the Seabees had built, after the Marines had taken a small chunk of land on the beach. As I ate the beans, I glanced over at row after row of white crosses, too far away and too dark to read the names. But I didn't have to. I knew that each cross marked the final resting place of some Marine who had gone as far as he was able in this mortal world of ours.

Before taking off everything seemed to be wrong that morning. My plane wasn't ready and I had to switch to another. At last minute the ground crew got my original plane in order and I scampered back into that. I was to lead a fighter sweep over Rabaul, meaning two hundred miles over enemy waters and territory again. We coasted over at about twenty thousand feet to Rabaul. A few hazy cloud banks were hanging around - not much different from a lot of other days. The fellow flying on my wing was Captain George Ashmun from New York City. He had told me before the mission: 'You go ahead and shoot all you want, Gramps. All I'll do is keep them off your tail.'

This boy was another who wanted me to beat that record, and was offering to stick his neck way out in the bargain. I spotted a few planes coming through the loosely scattered clouds and signaled to the pilots in back of me: 'Go down and get to work.' George and I dove first. I poured a long burst into the first enemy plane that approached, and a fraction of a second later saw the Nip pilot catapult out and the plane itself break out into fire. George screamed out over the radio: 'Gramps, you got a flamer!'

Then he and I went down lower into the fight after the rest of the enemy planes. We figured that the whole pack of our planes was going to follow us down, but the clouds must have obscured their view. Anyway, George and I were not paying too much attention, just figuring that the rest of the boys would be with us in a few seconds, as was usually the case. Finding approximately ten enemy planes, George and I commenced firing. What we saw coming from above we thought were our own planes - but they were not. We were being jumped by about twenty planes. George and I scissored in the conventional Thach weave way, protecting each other's blank spots, the rear ends of our fighters. In doing this I saw George shoot a burst into a plane and it turned away from us plunging downward, all on fire. A second later I did the same thing to another plane. But it was then that I saw George's plane start to throw smoke, and down he went in a half glide. I sensed something was horribly wrong with him. I screamed at him: 'For God's sake, George, dive!'

Our planes could dive away from practically anything the Nips had out there at the time, except perhaps a Tony. But apparently George had never heard me or could do nothing about it if he had. He just kept going down in a half glide. Time and time again I screamed at him, but he didn't even flutter an aileron in answer to me.

I climbed in behind the Nip planes that were plugging at him on the way down to the water. There were so many of them I wasn't even bothering to use my electric gun sight consciously, but continued to seesaw back and forth on my rudder pedals, trying to spray them all in general, trying to get them off George to give him a chance to bail out or dive - or do something at least. But the same thing that was happening to him was now happening to me. I could feel the impact of enemy fire against my armor plate, behind my back, like hail on a tin roof. I could see the enemy shots progressing along my wing tips, making patterns.

George's plane burst into flames and a moment later crashed into the water. At that point there was nothing left for me to do. I had done everything I could. I decided to get the hell away from the Nips. I threw everything in the cockpit all the way forward - this means full speed ahead - and nosed my plane over to pick up extra speed until I was forced by water to level off. I had gone practically a half a mile at a speed of about four hundred knots, when all of a sudden my main gas tank went up in flames in front of my very eyes. The sensation was much the same as opening the door of a furnace and sticking one's head into the thing.

Though I was about a hundred feet off the water, I didn't have a chance of trying to gain altitude. I was fully aware that if I tried to gain altitude for a bail-out I would be fried in a few more seconds."

Boyington landed in the water, badly injured. After being strafed by the Jap fighters, he struggled onto his raft until captured by a Jap submarine several hours later. They took him first to Rabaul, where he was brutally interrogated. Even the general commanding Japanese forces at Rabaul interviewed him. Pappy related in *Baa Baa Black Sheep* that the general asked him who had started the war. After Pappy replied that of course the Japanese had started the war by attacking Pearl Harbor, the general then told him this short fable:

"Once upon there was a little of old lady and she traded with five merchants. She always paid her bills, and got along fine. Finally the five merchants got together, and they jacked up their prices so high the little old lady couldn't afford to live any longer. That's the end of the story."

The general left the room, leaving Boyington to ponder that there had to be two sides to everything.

After about six weeks the Japanese flew him to Truk, where he experienced one of the early carrier strikes against that island in February of 1944. Along with six other captured Americans, he was confined in a small but sturdy wooden cell - which might have been designed for one inmate. The only opening was a six-inch hole in the floor for relieving themselves. With six men in a tiny cell this was unpleasant enough, but when the Japs actually overfed them with rice balls and pickles diarrhea resulted, and then the situation became really messy.

He eventually moved to a prison camp at Ofuna, outside of Yokohama. His autobiography relates the frequent beatings, interrogations, and near starvation that he endured for the next eighteen months. The guards, whose only qualification seemed to be passing "a minus-one-hundred I.Q. test," beat the prisoners severely for any infraction, real or imagined.

Boyington lost about eighty pounds, and described how he once entirely consumed a "soup bone the size of my fist" in just two days, a feat which previously he would not have believed a dog could achieve. During the middle period of his captivity he had the good fortune to be assigned kitchen duty. Here, a Japanese grandmother who worked in the kitchen befriended Greg and helped him filch food. Before long, he returned to his pre-captivity weight. He even got drunk on New Year's Eve, begging a little sake from each of the officers. From Camp Ofuna, he witnessed the first B-29 raids striking the nearby naval base at Yokohama.

When he was repatriated, Boyington found out he had been awarded the Medal of Honor and the Navy Cross. He also added to his claims for aerial victories after his return.

Several other pilots had seen him down one Zero, which raised his total to twenty with the Black Sheep, and twenty-six if his six with the Flying Tigers were included. Twenty-six was Eddie Rickenbacker's WWI record, and also the number shot down by Joe Foss, the top-scoring Marine pilot of all time. When the final two kills he claimed before being shot down were added, the total rose to twenty-eight.

Pappy lived until 1988 but it was a hard life marked by financial instability, marriages, divorces and battles with alcoholism. However it has been noted by many that, whatever his problems, he never seemed to lack for attractive female companionship. Things started downhill on his War Bond tour, when he was frequently drunk, and the Marine Corps placed Boyington on the retired list in 1947, allegedly for medical reasons.

Of course Pappy's greatest fame came in the mid Seventies, when the television show *Baa Baa Black Sheep* debuted. Based very loosely on Boyington's memoirs, the show had a three-year run and achieved a consistent popularity in re-runs. Pappy was a consultant to the show and got on well with its star, Robert Conrad - but the show's description of the Black Sheep pilots as a bunch of misfits and drunks, which Pappy happily went along with, destroyed his friendship with many of the squadron veterans. The show made Pappy a real celebrity, and along with his fourth wife Jo he made a good career out of being an entertainer - appearing at air shows, on TV programs, etc. Finally, after a long battle with cancer, Pappy died in 1988 - and so ended one of the most colorful lives any of us could hope to live.

UNCOMMON VALOR

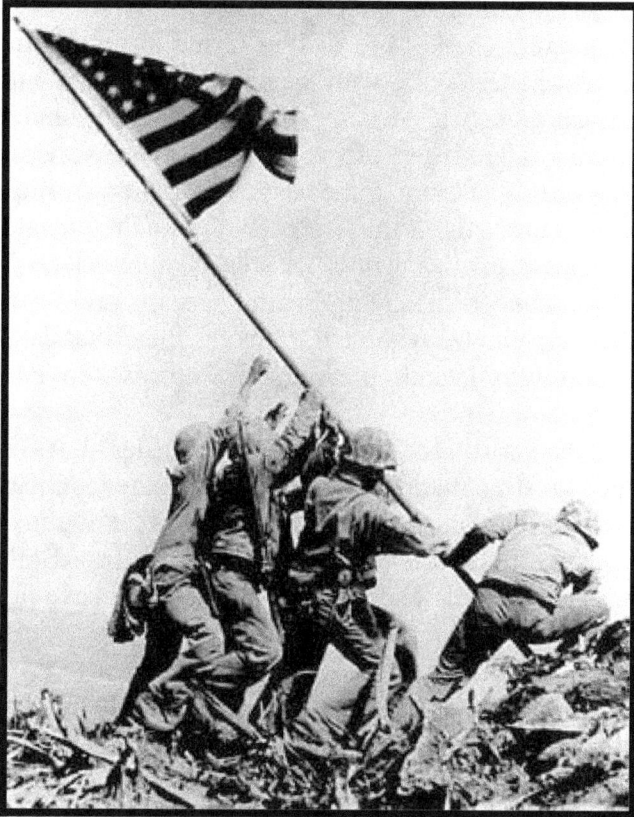

Iwo Jima

"Among the Americans who fought on Iwo Island, uncommon valor was a common virtue." – Admiral Chester Nimitz

Iwo Jima, which means Sulfur Island, was strategically important as an air base for fighter escorts supporting long-range bombing missions against mainland Japan. Because of the distance between mainland Japan and U.S. bases in the Mariana Islands, the capture of Iwo Jima would provide an emergency landing strip for crippled B-29s returning from bombing runs. The seizure of Iwo would also allow for sea and air blockades, along with the ability to conduct intensive air bombardment to destroy the enemy's air and naval capabilities. The seizure of Iwo Jima was deemed necessary, but the prize would not come easy. The fighting which took place during the thirty-six-day assault would be immortalized in the words of the Commander, Pacific Fleet/Commander in Chief, Pacific Ocean Areas Admiral Chester W. Nimitz, who said, "among the Americans who served on Iwo Island, uncommon valor was a common virtue."

Japan had fortified this island for decades, but in 1944 decided to strengthen it further. Under the command of General Kuribayashi, some 22,000 troops were dug in deeply and prepared to defend the island to the last man. Both ends of the island were defended in such a way as to have fields of fire in every direction. The morale of the Japanese defenders was high, and they were prepared to "give their lives for the Emperor." The Marines would accommodate them.

Initial carrier raids against Iwo Jima began in June of 1944. Prior to the invasion, the eight-square-mile island would suffer the longest, most intensive shelling of any Pacific island during the war. The 7th Air Force, working out of the Marianas, supplied B-24 heavy bombers for the campaign. In addition to the air assaults on Iwo, the Marines requested ten days of pre-invasion naval bombardment. Due to other operational commitments and the fact that a prolonged air assault had been waged on Iwo Jima, Navy planners authorized only three days of naval bombardment.

Unfavorable weather conditions would further hamper the effects.

Despite this, Admiral Turner decided to keep the invasion date as planned, and the Marines prepared for D-Day on February19th. More than 450 ships massed off Iwo as the H-Hour bombardment pounded the island. Shortly after 9 AM, Marines of the 4th and 5th Divisions hit beaches Green, Red, Yellow and Blue, initially finding little enemy resistance. Coarse black volcanic sand hampered the movement of men and machines as they struggled to move up the beach. As the protective naval gunfire subsided to allow for the Marine advance, the Japanese emerged from their fortified underground positions to begin a heavy barrage of fire against the invading force.

The 4th Marine Division pushed forward against heavy opposition to take the Quarry, a Japanese strong point. The 5th Marine Division's 28th Marines had the mission of isolating Mount Suribachi. The next day the climb up Suribachi started. The Japanese were dug in at hundreds of spots along the slope and had to be killed as the Marines went up. The top was an observation post from which the Japanese were directing fire on the Americans throughout the island. The Marines fought hand-to-hand and face-to-face with the defenders all day and into the next. There were at least two thousand Japanese dug in on the mount. The third day was even worse than the first two, if that was possible. It was on this day that a platoon of Marines became the most decorated platoon in Marine Corps history. The 3rd platoon of E company, 2nd Battalion, 28th Marine Regiment, 5th Marine Division received the following decorations from that battle - one Medal of Honor (posthumously to PFC Don Ruhl), two Navy Crosses, one Silver Star, seven Bronze Stars, and seventeen Purple Hearts. On February 20th, one day after the landing, the 28th Marines secured the southern end of Iwo and moved to take the summit of Suribachi. By day's end, one third of the island and Motoyama Airfield No. 1 were controlled by the Marines.

At 8 AM on February 23rd a patrol of forty men from 3rd Platoon, E Company, 2nd Battalion, 28th Marines, led by First Lieutenant Harold G. Schrier, assembled at the base of Mount Suribachi. The platoon's mission was to take the crater at Suribachi's peak and raise the U.S. flag. The Marines slowly climbed the steep trails to the summit, but encountered no enemy fire. As they reached the top, the patrol members took positions around the crater watching for pockets of enemy resistance as other members of the patrol looked for something on which to raise the flag. At 10:20 AM, the flag was hoisted on a steel pipe above the island. This symbol of victory sent a wave of strength to the battle-weary fighting men below, and struck a further mental blow against the island's defenders.

A Marine Corps combat photographer, Sergeant Lou Lowery, captured this first flag raising on film just as the enemy hurled a grenade in his direction. Dodging the grenade, Lowery hurled his body over the edge of the crater and tumbled fifty feet. His camera lens was shattered, but he and his film were safe.

Three hours later another patrol was dispatched to raise another, larger flag. The battle for Iwo Jima is encapsulated by this historic flag raising atop Suribachi, which was captured on film by Associated Press photographer Joe Rosenthal. His photograph, seen around the world as a symbol of American values, would earn him many awards including the 1945 Pulitzer Prize.

The 3rd Marine Division joined the fighting on the fifth day of the battle. These Marines immediately began the mission of securing the center sector of the island. Each division fought hard to gain ground against determined Japanese defenders. The Japanese leaders knew with the fall of Suribachi and the capture of the airfields that the Marine advance on the island could not be stopped, however they were determined to make the Americans fight for every inch of land.

Lieutenant General Tadamishi Kuribayashi, commander of the ground forces on Iwo Jima, concentrated his energies and his forces in the central and northern sections of the island. Miles of interlocking caves, concrete blockhouses and pillboxes proved to be one of the most impenetrable defenses encountered by the Marines in the Pacific.

The Marines worked to drive the enemy from the high ground. Their goal was to capture the area that appropriately became known as the "Meat Grinder." This section of the island included the highest point on the northern portion of the island, Hill 382, an elevation known as the "Turkey Knob," which had been reinforced with concrete and was home to a large enemy communications center, and the "Amphitheater," a southeastern extension of Hill 382.

The 3rd Marine Division encountered the most heavily fortified portion of the island in their move to take Airfield No. 2. As with most of the fighting on Iwo Jima, frontal assault was the method used to gain each inch of ground. By nightfall on March 9th the 3rd Division had reached the island's northeastern beach, cutting the enemy defenses in two.

On the left flank of the 3rd Marine Division, the 5th Marine Division pushed up the western coast of Iwo Jima from the central airfield to the island's northern tip. Moving to seize and hold the eastern portion of the island, the 4th Marine Division encountered a "mini banzai" attack from the final members of the Japanese Navy serving on Iwo. This attack resulted in the death of nearly seven hundred enemy and ended the centralized resistance of enemy forces in the 4th Division's sector. Shortly thereafter, a proud moment came for those who had worked so hard to gain control of the island when the first emergency landing was made by a B-29 bomber on March 4th.

Operations entered the final phases on March 11th, since enemy resistance was no longer centralized. Individual pockets of resistance were taken out one by one. The island was pronounced secured on the 14th of March, although the Marines would suffer six thousand more casualties cleaning up this "secured" island. The last Japanese attack came on the 26th of March when one thousand Japanese charged the American line. The Marines suffered more KIAs, but killed the Jap attackers almost to a man. The U.S. Army's 147th Infantry Division then assumed ground control of the island on April 4th, relieving the largest body of Marines ever committed to combat in one operation.

The thirty-six-day assault resulted in more than twenty-six thousand American casualties, including 6,800 dead. Of the twenty thousand Japanese defenders only 1,083 survived, and General Kuribayashi committed suicide in disgrace. The Navy suffered also. Five ships were sunk by Kamikaze planes, as was the carrier *HMS Bismarck* with five hundred of her crew. The Marines' efforts, however, provided a vital link in the U.S. chain of bomber bases. By the war's end 2,400 B-29 bombers carrying twenty-seven thousand crewmen made unscheduled landings on the island.

Historians described U.S. forces' attack against the Japanese defense as "throwing human flesh against reinforced concrete." Twenty-seven Medals of Honor were awarded to Marines and sailors, many posthumously - more than were awarded for any other single operation during the war. Over the years, the flag raising has come to symbolize the spirit of the Corps to all Marines. On November 10, 1954, a bronze monument of the flag raising, sculpted by Felix de Weldon and located in Arlington National Cemetery, was dedicated to all Marines who have given their lives in defense of their country. Then-Vice President Richard M. Nixon said:

"This statue symbolizes the hopes and dreams of America, and the real purpose of our foreign policy. We realize that to retain freedom for ourselves, we must be concerned when people in other parts of the world may lose theirs. There is no greater challenge to statesmanship than to find a way that such sacrifices as this statue represents are not necessary in the future, and to build the kind of world in which people can be free, in which nations can be independent, and in which people can live together in peace and friendship."

MANILA JOHN

Gunnery Sergeant John Basilone

"He contributed materially to the defeat, and virtually the annihilation, of a Japanese regiment." – Chesty Puller

John Basilone was born in 1916 in Buffalo, New York, one of ten children of Salvatore and Dora Basilone. Reared and educated in Raritan, New Jersey, John gained local attention as a light-heavyweight boxer. He enlisted in the Army when he was eighteen and served in the Philippines, where he picked up the nickname "Manila John." He was honorably discharged in 1937 but, anticipating World War II, he enlisted in the Marines in July of 1940.

John Basilone never cared much for the fame that accompanied his Medal of Honor - the parades, the newsreel appearances, the starlets who hung on his arm. He would much rather, he insisted, be just a "plain Marine" like his buddies who were still out in the Pacific. Eventually he got his wish, and became a legend known to every boot camp grad of the past forty-five years, including those now on duty in the Persian Gulf. When the USMC began the long, bloody process of rooting the Japanese from the islands of the Pacific, Basilone was front and center.

Guadalcanal was a fierce clash of national wills. Bloodied and humiliated by the sneak attack on Pearl Harbor, American armed forces were on the comeback trail less than six months after the debacle. At Guadalcanal, a disease infested island, two superb military organizations met each other for the first time in land combat - bayonet to bayonet - in a contest only one army could win.

The United States Marines were determined to keep their small foothold at Henderson Field, and the Japanese were equally determined to drive them into the sea. During the protracted battle which lasted for six months, the struggle to "own" Henderson Field came to a bloody climax a Sunday night in October of 1942.

At Lunga Ridge, which is about one thousand yards south of Henderson Field, it was raining torrents, creating miserable, bottomless mudholes - typical Guadalcanal weather. The Marines manning the main line of defense were exhausted. For two days Japanese human wave assaults had been flung against them. Each time the charging enemy had been driven off - but the weary Marines knew their tough adversaries weren't through. The Japanese would gather reinforcements and return.

About midnight, from the out of the ink-black darkness, came hundreds of screaming Japanese troops. Throwing themselves on the flesh-cutting barbed wire, the first of the waves formed human bridges for their comrades to leap across. One of the Marine section leaders facing them was Sergeant John Basilone. An experienced machine gunner, Basilone knew his guns would be tested to their mechanical limits - and it would be up to him to keep them firing.

During the attack, while grenades, small arms and machine guns were ripping through the night and exploding human flesh, splattering friend and foe, Sergeant Basilone stayed with his malaria-ridden men. Repeatedly repairing guns and changing barrels in almost total darkness, he ran for ammo or steadied his men who were firing full trigger to keep a sheet of white-hot lead pouring into the ranks of the charging Japanese.

Bodies piled so high in front of his weapons pits that the guns had to be reset so the barrels could fire over the piles of corpses. Not even the famous water-cooled heavy machine guns could stop all the assaults, and one section of guns was overrun. Two men were killed, and three others wounded.

Basilone took one of his guns on his back and raced for the breach in the line, where eight Japanese were surprised and killed. Basilone lifted a machine gun and its tripod - ninety pounds of weaponry in all - and raced two hundred yards to the silenced gun pit and started firing. Japanese bodies began stacking up in front of the emplacement. Enemy soldiers attacked his rear, and he cut them down with his pistol. Short of shells, he dashed two hundred yards amid a stream of bullets to an ammunition dump and returned with an armload of ammo. Flares lit up more swarms of grenade-tossing attackers. Basilone fired till heat blistered his hands, and still kept shooting. At dawn, he rested his head on the edge of the pit. Nearby lay thirty-eight enemy bodies. The line had held.

The next day the guns were jammed by mud and water, and a few yards away the Japanese were forming for another charge. Frantically stripping mud from the ammo belts, men fed them into the guns as Basilone cleared jams and sprayed the fiendish troops rushing at his positions with razor sharp bayonets and hands full of grenades.

Sometime after 0200 the firing finally died down, but no one relaxed, and at 0300 the officers of the Sendai Regiment prepared the remnants of their unit for a final Banzai charge. The full weight of the fanatical Japanese attack seemed to fall on Basilone's men, but he had set up a crossfire which smashed the charge. Dropping to the mud and still screaming, Colonel Sendai's remnants crawled forward trying to reach their tormentors. By depressing the muzzles of his weapons, Basilone destroyed them too.

Nash Phillips lost a hand fighting next to his Sergeant, and was surprised to see Basilone appear next to his bed a little while after dawn. "He was barefoot and his eyes were red as fire. His face was dirty black from gunfire and lack of sleep. His shirt sleeves were rolled up to his shoulders. He had a .45 tucked into the waistband of his trousers. He'd just dropped by to see how I was making out – me, and the others in the section. I'll never forget him. He'll never be dead in my mind!"

At dawn the battlefield was strewn with dead and wounded, but Henderson Field still belonged to the Americans and its ownership would never be seriously challenged again. At least thirty-eight dead Japanese were credited to Sergeant Basilone - and many of those were killed with his Colt .45 at almost arms length. Just twenty-six years old, Manila John Basilone had entered the ranks of the Marine Corps' pantheon of heroes - and shortly America would take the big, handsome Marine with jug ears and a smile like a neon sign to their hearts. The legend of a "Fighting Sergeant" was born.

When the battle was over and his squad members interviewed, Sergeant Basilone was credited by his men for his will to fight and ability to inspire them on a night of cold fear none would ever forget.

Within a short time the Japanese evacuated Guadalcanal, and prepared to meet other Marine invasions of their strongholds elsewhere in the Pacific. American fighting men had proven they could beat the best of the best, the most experienced troops Japan could throw at them. After Guadalcanal the Japanese high command had a fresh respect for the Marines, and they would be forced to meet them time and again as America pressed across the Pacific toward their homeland.

When he received the nation's highest decoration, John Basilone replied modestly, "Only part of this medal belongs to me. Pieces of it belong to the boys who are still on Guadalcanal. It was rough as hell down there." On the 1943 War Bond Tour Basilone was to say, "Doing a stateside tour is tougher than fighting Japs."

"Manila John" blushed when photographers snapped his picture while being kissed by a Hollywood starlet, smiled broadly when an oil portrait was unveiled in the tiny brick town hall, and was shyly grateful for the five thousand dollar war bond neighbors gave him. He also turned down the bars of a second lieutenant. "I'm just a plain Marine," he said, "and I want to stay one."

To millions Basilone was a hero, one of the first of the war, and he could have remained stateside training troops and selling war bonds. Instead he said farewell to his new wife, also a Marine, and joined the Fifth Division. Staying behind, he told his buddies, would be "like being a museum piece." And it wouldn't seem right, he said "if the Marines made a landing on the Manila waterfront and "Manila John" wasn't among them."

And so it was that Manila John Basilone landed on the black sandy beaches of Iwo Jima just months later. With the invasion ninety minutes old, the intrepid sergeant had one thought. "C'mon, you guys! Let's get these guns off the beach!" he yelled at the gunners just behind, backs hunkered low and straining under the heavy loads of weapons and ammunition amid the blistering fire. While under heavy artillery fire on February 19th, 1944 he single-handedly took out an enemy blockhouse - just before the wasplike whir of an incoming mortar sounded its eerie warning, and there was a shattering blast.

Basilone lunged forward in midstride, arms flung outward over his head. He and four comrades died in that instant. On his outstretched left arm was a tattoo which said, "Death before Dishonor!" Manila John wouldn't live to see Dewey Boulevard again, but he had won the Navy Cross, the Marine Corps' second highest decoration for valor.

VICTORY AT HIGH TIDE

Inchon

"The Marines and the Navy have never shone more brightly than this morning." - General Douglas MacArthur, Inchon,

It had been a long day filled with the sounds of battle, as thousands of frightened young Marines crowded their landing craft. Only five years earlier, most of them had been students finishing high school and learning world geography from the accounts of veterans of the world war in Europe and the Pacific. Despite the broad expanse of the World War II Theater, most of these young men had never heard of the

Asian peninsula of Korea – and none could probably find it on a map.

Still ringing in the ears of the young Marines of the 1st Marine Division were the words of the legendary Leatherneck commander of the first regiment, Colonel Lewis Chesty Puller. "We're the most fortunate of men. Most times, professional soldiers have to wait twenty-five years or more for a war, but here we are, with only five years wait for this one... we live by the sword, and if necessary we'll be ready to die by the sword. Good luck. I'll see you ashore."

The day was almost gone as the landing craft struggled against the treacherous tides to make their way to the shore at Inchon. Wooden scaling ladders protruded from the front of the low profile barges that transported the Marines towards their destination. It was an amphibious assault against three enemies - the soldiers of North Korea, the quickly fading daylight hours, and the infamously dangerous geography of Korea's west coast. Indeed, the 1st Marine Division commander, Major General Oliver P. Smith, had noted, "Half the problem was getting to Inchon at all."

General Douglas MacArthur's most ardent detractors will admit that the surprise amphibious assault at Inchon, dubbed *Operation Chromite*, was a stroke of military genius. In a matter of days the highly successful operation broke the back of the North Korean invasion of the South, and liberated the capitol city of Seoul.

Prior to World War II the Asian peninsula of Korea was undivided, first as an independent kingdom, then as a Protectorate of Japan (1910-45). Then, shortly before World War II came to a close, the United States and Russia reached an agreement to divide the peninsula at the 38th parallel for the purpose of accepting the surrender of Japanese troops. When war ended both nations worked hard to promote friendly governments, with Russia suppressing the moderate nationalists and supporting Kim Il Sung in the North, and the United States promoting United Nations supervised elections in the South. These elections led to the formation of the Republic of South Korea (ROK) in August of 1948. The following month the inhabitants north of the 38th parallel responded by establishing the People's Republic of Korea (DPRK). For the first time in history, the peninsula was divided into North and South Korea.

On Sunday, June 25th, 1950 the North Koreans attempted to reunite the two nations of the Asian peninsula. At exactly 4 A.M. nearly 100,000 DPRK soldiers, supported by tanks and 130 aircraft, attacked across the border. Three days later the capitol of Seoul, just fifty miles south of the border between the two countries, had fallen to the North, and within weeks ROK, U.S., and U.N. forces had been pushed all the way back to Pusan on the southeast coast. The NKPA (North Korea People's Army) held most of the peninsula and appeared close to uniting their land under the banner of Communism.

Throughout the months of July and August the United States moved quickly to shore up defenses at Pusan with supplies and an infusion of new troops. Throughout the period, from his headquarters in Japan, General Douglas MacArthur continued to hammer out the details for making a counterattack, beginning with an invasion at Inchon.

Actually, *Operation Chromite* was planned and proposed in early July when the war in Korea was barely a week old. It was typically MacArthuresque, transporting a large force completely around the enemy to land behind them, thus blocking supply routes and cutting off any retreat. The harbor at Inchon afforded all strategic requirements. It was located almost directly opposite Pusan, far to the enemy's rear flank. A successful invasion would cut off the NKPA from their command in the north, as well as their supply routes. According to military intelligence reports, the harbor was only lightly defended. The NKPA had committed the bulk of their invading force, some seven full divisions, to the effort at Pusan. Inchon was also located only ten miles from Seoul, the South Korean capitol which was now under enemy control.

The Joint Chiefs of Staff approved MacArthur's planned invasion early in the war. The course of events in and around Pusan delayed implementation, and changed the schematics of what was originally planned to commence on July 22nd with an assault by the 1st U.S. Cavalry. The 1st Cavalry was thrown instead into Korea east of Taejon, and General MacArthur turned his attention to the 1st and 5th Regiments of the 1st Marine Division to lead the Inchon invasion, along with the men of MacArthur's sole reserve unit in Japan, the Army's 7th Infantry Division.

D-Day was September 15th, as nearly seventy thousand American soldiers and Marines approached Inchon in a task force of 320 warships supported by four air craft carriers. At 5 AM Marine Corsairs struck the small island of Wolmi-do, followed within an hour by the initial Marine landing. Half an hour later the small island at the approach to Inchon was under American control, 108 enemy had been killed, and 136 captured. Marine casualties were light, with only seventeen Americans wounded.

While the island was the focus of the initial assault, the bulk of the X Corps assault force pulled back into the deeper waters of the Kanghwa Bay. The primary assault on Inchon itself would be much more difficult. General O. P. Smith had been more than astute in his observation that "Half the problem was getting to Inchon at all." Despite all of the tactical advantages Inchon posed for an amphibious assault to turn the tide of war in Korea, all of the geographical characteristics were negative.

The city of Seoul sits on the Han River, which runs northwest to spill into Kanghwa Bay and the Yellow Sea. The infamous coastal tides are among the most dangerous in the world, and have caused sand carried by the Han and numerous smaller rivers to create large beds of soft mud. When the high tides are running the swift currents of two to three knots - and sometimes up to ten knots - make navigation extremely dangerous. When the tides recede, hundreds of yards of mud flats extend outward from the shoreline. An invading force approaching from the sea could quickly sink up to its knees while it struggled to gain the firmer ground of the peninsula.

The men of the Third Battalion, 5th Marines that landed at Wolmi-do came in with the high tide, with a tide range of thirty-six feet. When the tide withdrew the island was surrounded by a sea of mud, separating the American force from the mainland as well as the rest of the assault force. The main assault was planned to return with the high tide nearly twelve hours later to land on the mainland itself. The timing meant that their small landing craft would have to struggle against the currents, negotiate the treacherous rocks and mud flats, reach the shoreline, and disgorge the Marines. Upon landing, the Marines would face a sixteen-foot sea wall, which they planned to scale with the ladders carried in each LST. There would be only about two hours of daylight to reach the shore, scale the walls, and set up their defensive positions. It was a formidable, frightening task for young men unaccustomed to war, and was made worse by the fact that the soldiers of the NKPA knew the Americans were coming.

Riding the crest of the incoming tide, the ships of the American task force carefully began their second approach to the harbor at Inchon on September 15th, 1950. Battleships filled the air with a hail of rockets, and explosions erupted all along the Korean shoreline. At 3:35 PM the men of the 1st Marine Division began loading in their landing craft. The LCVPs each carried twenty-two men and the needed scaling ladders. These shallow draft flat-bottom boats were well suited for shallow waters. Most would come in with the tide, unload at the shoreline, and then remain beached throughout the night as the tide withdrew.

The Marines planned to strike at two locations, with the remaining two battalions of the 5th Marine Regiment unloading at Red Beach while the entire 1st Regiment would forge its way across two miles of mud flats covered by shallow water to land at Blue Beach.

At 5:33 PM the first landing craft reached Red Beach and dropped its ramp, and determined Marines quickly lobbed grenades over the sea wall to discourage any enemy soldiers awaiting their arrival. When the scaling ladders were in place the assault began. From the rear of one of the landing craft a photographer snapped a picture. Leading the way, with only his back visible to the camera, was twenty-five-year-old Marine Corps First Lieutenant Baldomero "Punchy" Lopez. The young officer from Tampa, Florida would not only command his Marines into the foray... he would LEAD them.

All along Red Beach the Marines scaled the walls and were met with a tremendous volley of fire from the enemy. Lieutenant Lopez led his Third Platoon of Company A towards a nearby trench, killing a dozen North Korean soldiers in the process. During the opening ten minutes of the invasion however, eight Marines were also killed.

Lieutenant Lopez noted the heaviest enemy fire was coming from two nearby bunkers. Quickly he destroyed the first, and then ordered his men to provide covering fire while he crawled towards the second. Nearing the enemy position, the brave lieutenant rose to throw a grenade. A sudden burst of automatic fire raked Baldomero's body, shattering his right arm and puncturing his chest. Thrown backward by the force of the enemy bullets, the armed grenade fell from his shattered hand.

Fighting intense pain and weak from loss of blood, Lieutenant Lopez dragged his body forward in an effort to retrieve and throw the grenade with his remaining good arm. He was unable to grasp it firmly and realized it would detonate within seconds, killing or wounding some of his nearby Marines. Unwilling to risk their lives, Lieutenant Baldomero Lopez pulled the grenade into the crook of his good arm and rolled over on it, absorbing the full impact of the explosion. He was instantly killed, but his Marines were saved.

General MacArthur later referred to the landing at Inchon as one of the Marine Corps' "finest hours." By nightfall most of the major objectives had been achieved, and by the following morning Inchon was secure. More than three hundred enemy soldiers had been killed, and nearly 1,500 wounded. The Marines lost twenty men killed in action, and 187 wounded. Twelve days later Gunnery Sergeant Harold Beaver ripped down the North Korean flag and raised the Stars and Stripes over the Government House in Seoul.

The shattered body of Lieutenant Lopez was returned to his hometown of Tampa for burial, and less than a year after his death, on August 30th, 1951, his parents were invited to the Pentagon where Secretary of the Navy Dan A. Kimball presented them with the Medal of Honor in recognition of their son's heroism and sacrifice.

First Lieutenant Lopez was the first of forty-two Marines to receive the Medal of Honor during the Korean War of 1950-1953. He was not, however, the first Marine to earn the Medal of Honor for heroism in Korea.

Little known and rarely remembered by most Americans, was the amphibious assault United States Marines and Navy Bluejackets had made in these same waters nearly seventy-five years earlier. Barely ten miles from where Lieutenant Lopez had led his men into battle and sacrificed his life, a young naval lieutenant had similarly led his men into armed combat. In the battle that followed, six Marines and nine sailors earned the first Medals of Honor to be awarded for foreign service. It was the Korean campaign of 1871, known to the Koreans as Shinmiyangyo. But that is another story...

FROZEN CHOSIN

The Chosin Reservoir

"The American First Marine Division has the highest combat effectiveness in the American armed forces. It seems not enough for our four divisions to surround and annihilate its two regiments. (You) should have one or two more divisions as a reserve force." – Chairman Mao Zedong's orders to Chinese General Song Shilun

It was Thanksgiving Day of 1950 at North Korea's Chosin Reservoir, and nighttime temperatures plunged to thirty degrees below zero. The ground was frozen solid. Night fell at 4:30 PM, and light did not return for nearly sixteen hours. This was an inhospitable place, even for the battle-tested men of the 1st Marine Division and the Army's 7th Infantry

Division, some of whom had fought through the worst of World War II.

Five months earlier, on June 25th, the North Korean People's Army (NKPA) had invaded South Korea and shattered the five-year-old peace. President Truman's response was swift and decisive, as was that of the newly formed United Nations. U.S. air and sea assets were committed immediately, and ground troops were committed on June 30th. Army General Douglas MacArthur was put in charge of the U.N. Command, which included combat and medical units from twenty-two nations.

At first the NKPA moved down the Korean Peninsula with relative ease, but on September 15th MacArthur launched his brilliant amphibious landing of X Corps at Inchon, deep behind enemy lines. The landing of the 1st Marine Division opened the door for an allied victory, and the Army's 7th Infantry Division came ashore and fought beside the Marines to recapture Seoul. Within weeks, the North Koreans were pushed back across the 38th parallel.

Once there American and U.N. leadership, civilian and military alike, decided to keep fighting all the way to the Yalu River, North Korea's border with China, intending to destroy the NKPA and unify the two Koreas under South Korean President Syngman Rhee. The allies were on the offensive, and most believed they would be home by Christmas - but Chinese leaders, with a large standing army, warned more than once they would intervene if U.N. forces crossed the 38th parallel.

The U.S. military was not ready for a ground war. After World War II and the debut of the atomic bomb, the Army and Marine Corps were rapidly demobilized. Equipment budgets were slashed, and in its new role as a peacekeeping force the Army of June 1950 was ill-equipped, under strength, and poorly trained. At the same time the Marine Corps, despite suffering a similar lack of resources, had continued to train for combat.

As the Marine Corps and Army prepared to cross the 38th parallel, MacArthur ordered Lieutenant General Walton H. Walker's Eighth U.S. Army up the west side of the peninsula. MacArthur divided X Corps, commanded by Major General Edward M. Almond, by landing the 1st Marine Division (under Major General O.P. Smith) at Wonsan on October 26th and the Army's 7th Infantry Division at Iwon on October 29th. The Army's 7th Infantry Division was the least prepared for war, because it had been stripped of many experienced officers and NCOs to fill the three divisions that first deployed to Korea.

Although orders changed many times, the plan was for the Marines to attack from Yudam-ni at the Chosin Reservoir, moving north and west, and ultimately meet the Eighth U.S. Army and cut off the NKPA in a pincer movement. The 31st Regimental Combat Team (RCT), composed of elements of the 7th Division, would attack northward along the east side of the Chosin while the 3rd Infantry Division was to hold the areas of Wonson and Hungnam and keep the roads open.

These forces would not have communication with one another, however. X Corps and the Eighth U.S. Army had a mountain range between them, while the reservoir separated the Marines from the 31st RCT. MacArthur's commanders were outraged that the forces were divided - and therefore vulnerable - but their protests accomplished nothing.

As the allied forces moved north, the Chinese first hit them in early November. Aerial reconnaissance pilots reported Chinese forces massing on the Yalu, and by mid-November Chinese strength at the Yalu was estimated at 300,000 - but MacArthur discounted these reports.

These early battles were intense but brief, with the Chinese retreating into the hills as quickly as they appeared. The Chinese Communist Forces' (CCF's) first offensive tested allied capability and put the Eighth U.S. Army and X Corps in check until the Chinese were ready for a more massive engagement. This tactic of pulling back lured the Americans deeper into enemy territory - and time was on China's side. While American units moved through North Korea, a pleasant October autumn became an early, bitterly cold winter.

Uneasy about the Chinese threat, Smith moved the 1st Marine Division north carefully, keeping his units close together to avoid being separated. He stockpiled supplies and ammunition and stationed units along the division's main supply route (MSR) to keep it open. All but ignoring MacArthur's order for speed, Smith incensed Almond.

Colonel Lewis B. "Chesty" Puller, commander of the 1st Marines, would hold the MSR. His 1st Battalion held Chinhung-ni at the base of the Fuchilin Pass, the 2nd Battalion was with Puller at Koto-ri eleven miles up the road, and the 3rd Battalion would support Smith's headquarters at Hagaru-ri at the base of the reservoir.

On the east side of the reservoir the 5th Marines, under Lieutenant Colonel Raymond L. Murray, protected the Marines' right flank until they were relieved by Army Lieutenant Colonel Don Faith's 1st Battalion, 32nd Infantry. Before moving toward Yudam-ni to join the 7th Marines, Murray warned Faith about the enemy presence and advised him to keep his forces tight.

By November 27th more elements of the 31st RCT had arrived east of the reservoir. The 31st RCT was hastily thrown together, and was composed of whichever units could move to replace the 5th Marines soonest. These included the

1st Battalion of the 32nd Infantry, the 3rd Battalion of the 31st Infantry, the 57th Field Artillery Battalion, and the 31st Tank Company. The 2nd Battalion, 31st Infantry never made it to the reservoir.

Faith moved his men far forward to occupy the area left by the 5th Marines - an area too large for one battalion. The remainder of the 31st RCT set up a second perimeter to the south. Again, forces were divided with an enemy threat present. The perimeters were loose, but MacLean planned to attack first thing in the morning. Few seemed worried about the dangerous situation for one night. The men of the 31st RCT later said they didn't believe the warnings about the Chinese.

Meanwhile Smith continued north from Koto-ri and established his command post at Hagaru-ri. He ordered airstrips scratched from the frozen earth there and at Koto-ri. His solid protection of the MSR and the airstrips would prove crucial to the breakout of the Marines and soldiers.

Over in the west, unknown to X Corps, a massive force of eighteen Chinese divisions had attacked the Eighth U.S. Army on November 25th and nearly destroyed it. Within two days it was in full retreat, but for the moment MacArthur kept his commanders in the east in the dark.

On November 27th, the reunited 5th and 7th Marine Regiments began their attack north from Yudam-ni and quickly ran into enemy resistance. The 7th Marines' commander, Colonel Homer L. Litzenburg Jr., sent Fox 2/7 to hold the high ground at Toktong Pass. The subsequent success of the fighting withdrawal depended on the tenacity of the young company commander, Marine Captain William Barber, and his men holding this crucial piece of ground.

That night, as temperatures plunged well below zero in the rugged mountains of North Korea, three Chinese divisions sounded horns, whistles, and bugles and attacked the 5th and 7th Marine regiments at the reservoir. Smith's worst fears became reality. That same night MacLean's men were jarred awake by more noisemakers, as two Chinese divisions breached their perimeter. With all other officers in the area dead or wounded, Marine Captain Ed Stamford, a World War II veteran and pilot attached to Faith with a team of four Marines as his tactical air control party, took command of A Company. Though not an infantryman, he rallied the company to repel the attack.

When MacArthur got reports of the ferocious Chinese assault he decided that X Corps would withdraw to Hungnam while the weakened Eighth U.S. Army would try to hold P'yongyang. His late call proved fatal, and during the next two weeks Marines and soldiers fought day and night to break out of the trap the Chinese had set.

The CCF surrounded everybody - the 11th Marines, the division's artillery, and the 5th and 7th Marines at Yudam-ni, Fox Company at Toktong Pass, Smith and his men at Hagaru-ri, Puller at Koto-ri, and the 31st RCT east of Chosin. They attacked late at night and retreated to the mountains during the day when deadly American close-air support was on the scene. Forward air controllers like Stamford would direct these attacks with barely functioning radios. X Corps might have been lost but for Navy, Marine Corps, and Air Force pilots performing bombing runs, close-air support, supply and ammunition drops, and the evacuation of thousands of wounded.

The 31st RCT continued to take heavy fire, and casualties mounted. The unbearable cold and frostbite also took its toll. MacLean was injured, captured, and later reported dead, leaving Faith in charge. Though he was a World War II veteran with no combat experience, his men describe him today as a charismatic leader who worked hard to get everyone out alive.

After four nights and five days of mounting casualties with no relief or rescue in sight, Faith decided the 31st RCT would fight its way out. He radioed Smith at Hagaru-ri and asked for support. MacArthur and Barr also had talked to Smith about sending a team to rescue the 31st RCT, which had by this time become known as Task Force Faith, but Smith's situation was not much better. Under constant enemy attack, he had everyone, including cooks and engineers, on the line holding the perimeter. Diverting support to the east would probably spell the loss of Hagaru-ri, which in turn would mean the end of the 5th and 7th Marines. Faith was on his own.

Faith's plan was to move out as soon as air support was available on December 1st, but clouds kept the unit in place until around 1 PM, leaving less than four hours of daylight. The breakout moved quickly at first, then came under heavy fire and hit enemy roadblocks. Young officers pulled even younger soldiers together to continue the fight. NCOs like Corporal George Pryor (the units were so jumbled up, the men thought he was a captain) rallied soldiers looking for leaders. It was the only way they would get out alive.

Command and control were lost, and Task Force Faith was fighting its way out in small pieces. Lending to the confusion, communication was by voice only - Stamford had the only working radio, and his was feverishly calling for air strikes and support. Ammo was low. Pilots tried to resupply the column, but some air drops drifted over to the enemy. Bullets rained down on the column. Soldiers took cover and returned fire as best they could, but they were surrounded. Stamford continued to call in air strikes, with the enemy so close that some Americans were hit by napalm.

After about four miles, the column halted because the lead drivers were dead. Faith himself lay in a jeep dying, and his task force died with him. Organization broke down, and it was now every man for himself. The enemy continued to close and kept firing. Officers and soldiers grabbed what wounded they could and fought their way out of what had become a death trap. Some played dead and escaped later, and those who did not get out were killed or captured.

Many who made it out headed across the frozen, unprotected Chosin Reservoir. Over the next several days hundreds walked, crawled, or were dragged across the ice to the Marines' perimeter four and a half miles away at Hagaru-ri. A group of Marine volunteers and a Navy hospital corpsman led by Marine 1st Motor Transport Battalion Commander Lieutenant Colonel Olin Beall spent several days out on the ice and brought in about 320 soldiers in two days.

Some members of Task Force Faith made it to Hagaru-ri on their own. One survivor, 1st Lieutenant John Gray, remembers a vigilant and suspicious Marine at the perimeter asking him for the password. After days of combat on the other side of the reservoir, who knew? The sergeant then asked Gray the location of several cities such as Dubuque, Des Moines, and Sioux City. Luckily Gray knew his geography, and he and his men were welcomed into the perimeter.

A total of about 1,050 of 31st RCT's 2,500 men had survived. About 385 were considered able-bodied and fought at Hagaru-ri and all the way to the sea. Barr, devastated by the loss of his men, was relieved of command shortly after Chosin.

Meanwhile, Smith had heard about MacArthur's order to withdraw on November 30th and reportedly huffed, "It took them two days to decide this?" He ordered his 5th and 7th Marines to pull back to Hagaru-ri. This would not be easy, since they were still surrounded at Yudam-ni, and the MSR was interrupted and full of enemy soldiers.

A reporter with Smith in Hagaru-ri labeled the Marine operation a retreat. Smith patiently explained that because they were surrounded and there was no rear, "retreat" was inaccurate - they would have to fight their way out. People back home read, "Retreat, hell, we're just attacking in another direction." Though not in Smith's style, this was the perfect description of the Marines' problem and their solution, and he never denied the quote.

The reporters also wanted time with the legendary Puller, who obliged with a highly quotable assessment of the situation. "We've been looking for the enemy for several days now. We finally found them. We're surrounded. That simplifies our problem of finding these people and killing them."

Back at Yudam-ni, Murray and Litzenburg decided to move by road during the day. Daylight gave them the advantage of air and artillery support. During the days and nights of battle, Barber and his company were alone (except for the enemy) at Toktong Pass. For the movement south, the pass had to be held, and Marine Lieutenant Colonel Raymond G. Davis' 1st Battalion, 7th Marines was going to relieve Barber and secure it.

Davis and his men were the first unit out of Yudam-ni. They traveled over the rough, steep terrain in the dark, bitter cold - something the Chinese would not expect - and made it undetected by the enemy. Davis found that Barber and his men had held the pass for five days despite relentless attacks. Casualties were high - of two hundred men, twenty-six had been killed, eighty-nine wounded, and three were missing. Air drops of ammo had proved invaluable.

Once Davis' men secured Toktong Pass, the 5th and 7th fought their way to Hagaru-ri. It took them seventy-nine hours to travel fourteen miles carrying the wounded and most of their equipment, but on December 3rd they entered the Hagaru-ri perimeter. Prisoner-of-war interrogations - extremely reliable at this point in the war - indicated that there were at least seven CCF divisions near Hagaru-ri. The Chinese knew its strategic location was the key to the Marines' successfully breaking out.

Once in Hagaru-ri the 5th, 7th, and other units rested, regrouped, and prepared for their next move, south to Koto-ri. Air Force C-46s and C-47s and other U.N. aircraft began an evacuation of about 4,300 wounded and frostbite victims. Smith gave the dead priority, which again outraged Almond, though Smith was adamant that fallen Marines held a special place and would be flown out first. About 140 were flown to Japan, while more than five hundred replacement combat Marines were flown in.

On December 6th the men at Hagaru-ri began their nine-mile, thirty-eight--hour fight to Koto-ri. Despite CCF control of the road and many roadblocks, the lead units moved through and kept the road open for Hagaru-ri's rear guard. About ten thousand men and one thousand vehicles reached the relative safety of Koto-ri, and once within the perimeter most of the 1st Marine Division again was reunited. More wounded were evacuated from the Koto-ri airstrip, and X Corps prepared for the forty-three mile fight to the sea.

Chinese POWs revealed that Fuchilin Pass would be the site of a major enemy attack. A CCF division lay in wait, three other CCF divisions were in the area, and another two were to be held in reserve. Lieutenant Colonel Donald M.

Schmuck's 1st Battalion, 1st Marines, who had been holding Chinhung-ni, was rested and ready to go, and on the snowy night of December 8th they surprised the Chinese.

Fuchilin Pass was the enemy's last major offensive during the Chosin campaign. The CCF had overextended its supply lines, and its soldiers were suffering from the cold and lack of food. The enemy would continue to launch minor assaults, but they were minimal compared to the force with which the CCF had struck at the reservoir.

Smith and his men reached Hungnam on December 11th, and by the 15th Navy ships were transporting them south. Smith's insight and careful, deliberate style made him the ideal commanding general for Chosin, and he was fortunate to have talented, experienced leadership from Puller, Murray, Litzenburg, Davis, and others. While his men fought together like a machine, it was his uncommon understanding of the situation - and a certain amount of luck - that ensured the story of the Chosin Reservoir would become part of American military lore.

HOWARD'S HILL

Staff Sergeant James E. Howard

"Retrieving wounded comrades from the field of fire is a Marine Corps tradition more sacred than life." – Robert Pisor

The Marine Corps has a tested tradition - it will never leave alone on the field of combat one of its fighting men. It will go to fantastic lengths and commit to battle scores of men to aid and protect a few. This is the story of a few such Marines, of the battle they fought, and the help they received from all the services - not just the Marine Corps.

Some twenty miles inland to the west of the Marine base at Chulai runs a range of steep mountains and twisting valleys. In that bandit's lair, the Viet Cong and North Vietnamese could train and plan for attacks against the heavily populated seacoast hamlets, massing only when it was time to attack. In early June of 1966, the intelligence reports reaching III MAF headquarters indicated that a mixed force of Viet Cong and North Vietnamese was gathering by the thousands in those mountains. But the enemy leaders were not packing their troops into a few large, vulnerable assembly points. Instead they kept their units widely dispersed, moving mainly in squads and platoons.

To frustrate that scheme and keep the enemy off balance, the Marines launched Operation KANSAS, an imaginative concept in strategy. Rather than send full infantry battalions to beat the bushes in search of small enemy bands, Lieutenant General Lewis Walt detailed the reconnaissance battalion of the 1st Marine Division to scout the mountains. The reconnaissance Marines would move in small groups of eight to twenty men, and if they located a large enemy concentration Marine infantry would be flown in. If, as was expected, they saw only numerous small groups of Viet Cong and North Vietnamese, they were to smash them by calling in air and artillery strikes.

Lieutenant Colonel Arthur J. Sullivan had set high training standards for his battalion. Every man had received individual schooling in forward observer techniques and reconnaissance patrol procedures. He was confident his men could perform the mission successfully, despite the obvious hazards. "The Vietnam War," he said, "has given the small-unit leader - the corporal, the sergeant, the lieutenant - a chance to be independent. The senior officers just can't be out there looking over their shoulders. You have to have confidence in your junior officers and NCOs."

One such NCO was Staff Sergeant Jimmie Earl Howard, who was the acting commander of the 1st Platoon, Charlie Company, 1st Reconnaissance Battalion. A tall, well-built man in his mid-thirties, Howard had been a star football player and later a coach at the San Diego Recruit Depot. Leadership came naturally to him. "Howard was a very personable fellow," his company commander, Captain Tim Geraghty said. "The men liked him. They liked to work for him." In Korea he had been wounded three times and awarded the Silver Star for bravery. In Vietnam he would receive a fourth Purple Heart - and the Medal of Honor.

As dusk fell on the evening of 13 June, 1966 a flight of helicopters settled on the slope of Hill 488, twenty-five miles west of Chulai. Howard and his seventeen men jumped out and climbed the steep incline to the top. The hill, called Nui Vu, rose to a peak of nearly fifteen hundred feet and dominated the terrain for miles. Three narrow strips of level ground ran along the top for several hundred yards before falling abruptly away. Seen from the air, they roughly resembled the three blades of an airplane propeller. Howard chose the blade which pointed north for his command post and placed observation teams on the other two. It was an ideal vantage point.

The enemy knew it also. Their foxholes dotted the ground, each with a small shelter scooped out two feet under the surface. Howard permitted his men use of these one-man caves during the day to avoid the hot sun and enemy detection, since there was no other cover or concealment to be found. There were no trees, only knee-high grass and small scrub growth.

In the surrounding valleys and villages, there were many enemy. For the next two days, Howard was constantly calling for fire missions, as members of the platoon saw small enemy groups almost every hour. Not all of the requests for air support and artillery strikes were honored. Sullivan was concerned that the platoon's position, so salient and bare, might be spotted by a suspicious enemy. Most of the firing at targets located by the platoon was done only when there was an observation plane circling in the vicinity to decoy the enemy. After two days Sullivan and his executive officer, Major Allan Harris, became concerned with the risk involved in leaving the platoon stationary - but the observation post was ideal. Howard had encountered no difficulty, and in any case he thought he had a secure escape route along a ridge to the east. It was decided to leave the platoon on Nui Vu for one more day.

However, the enemy was well aware of the platoon's presence. Sullivan had a theory that the Viet Cong and North Vietnamese, long harassed, disrupted and punished by reconnaissance units in territory they claimed to control absolutely, had determined to eliminate one such unit, hoping thereby to demoralize the others. Looked at in hindsight, the ferocity and tenacity of the attack upon Nui Vu gives credence to that theory. In any case, the North Vietnamese made their preparations well and did not tip their hand. On 15 June they moved a fresh, well equipped and highly trained battalion to the base of Nui Vu, and in late afternoon hundreds of the enemy started to climb up the three blades, hoping to annihilate the dozen and a half Marines in one surprise attack.

The Army Special Forces frustrated that plan. Sergeant First Class Donald Reed and Specialist Hardey Drande were leading a platoon of CIDG (Civilian Irregular Defense Group) forces on patrol near Nui Vu that same afternoon. They saw elements of the North Vietnamese battalion moving towards the hill and radioed the news back to their base camp at Hoi An, several miles to the south. Howard's radio was purposely set on the same frequency, so he was alerted at the same time. Reed and Drande wanted to hit the enemy from the rear and disrupt them, but had to abandon the idea when they suddenly found themselves a very unpopular minority of two on the subject. Describing the reactions of the Special Forces NCOs later, Howard could not resist chuckling. "The language those sergeants used over the radio," he said, "when they realized they couldn't attack the NVA, well, they sure didn't learn it at communications school." Even though the Special Forces were not able to provide the ground support they wished to, their warning alerted Howard and enabled him to develop a precise defensive plan before the attack was launched.

Acting on the report, Howard gathered his team leaders, briefed them on the situation, selected an assembly point and instructed them to stay on full alert and withdraw to the main position at the first sign of an approaching enemy. The corporals and lance corporals crept back to their teams and briefed them in the growing dusk. The Marines then settled in to watch and wait.

Lance Corporal Ricardo Binns had placed his observation team on a slope forty meters forward of Howard's position. At approximately 2200 that evening, while the four Marines were lying in a shallow depression discussing in whispers their sergeant's solemn warnings, Binns quite casually propped himself up on his elbows and placed his rifle butt in his shoulder. Without saying a word, he pointed the barrel at a bush and fired. The bush pitched backward and fell thrashing twelve feet away.

The other Marines jumped up, and each threw a grenade before grabbing his rifle and scrambling up the hill. Behind them grenades burst and automatic weapons pounded away. The battle of Nui Vu was on.

The other outposts withdrew to the main position. The Marines commanded a tiny rock-strewn knoll. The rocks would provide some protection for the defenders. Placing his two radios behind a large boulder, Howard set up a tight circular perimeter, not over twenty meters in diameter, and selected a firing position for each Marine.

The North Vietnamese were setting up too. They had made no audible noises while climbing. There was no talking, no clumsy movements. When Binns killed one of their scouts, they were less than fifty meters from the top.

The Marines were surrounded. From all sides the enemy threw grenades. Some bounced off the rocks, some rolled back down the slopes, and some did not explode. But some landed right on Marines, and did explode. The next day the platoon corpsman, Billie Don Holmes, recalled, "They were within twenty feet of us. Suddenly there were grenades all over. Then people started hollering. It seemed everyone got hit at the same time."

Holmes crawled forward to help, and when a grenade exploded between him and a wounded man he lost consciousness.

The battle was going well for the NVA. Four .50-caliber machineguns were firing in support of the assault units, their heavy explosive projectiles arcing in from the four points of the compass. Tracer rounds from light machineguns streaked toward the Marine position, pointing the direction for reinforcements gathering in the valley. 60mm mortar shells smashed down and added rock splinters to the shrapnel whizzing through the air.

The NVA followed up the grenade shower with a full, well coordinated assault, directed by shrill whistles and the clacking of bamboo sticks. From different directions, they rushed the position at the same time, firing automatic weapons, throwing grenades, and screaming. Howard later said he hadn't been sure how his troops would react. They were young, and the situation looked hopeless. They had been shocked and confused by the ferocity of the attack and the screams of their own wounded.

But they reacted savagely. The first lines of enemy skirmishers were cut down seconds after they stood up and exposed themselves. The assault failed to gain momentum, and the North Vietnamese in the rearward ranks had more sense than to copy the mistakes of the dead. Having failed in their swift charge, they went to earth and probed the perimeter, seeking a weak spot through which they could drive. To do this, small bands of the enemy tried to crawl close to a Marine and then overwhelm him with a burst of fire and several grenades.

But the Marines too used grenades, and the American hand grenade contains twice the blast and shrapnel effect of the Chinese Communist stick grenade. The Marines could also throw farther and more accurately than the enemy. A Marine would listen for a movement, gauge the direction and distance, pull the pin, and throw. High-pitched howls and excited jabberings mingled with the blasts. The North Vietnamese pulled back to regroup.

Howard had taken the PRC-25 radio from one of his communicators, Corporal Robert Martin, and during the lull contacted Captain Geraghty and Lieutenant Colonel Sullivan. With his escape route cut off and his force facing overwhelming odds, Howard kept his message simple. "You gotta get us out of here," he said. "There are too many of them for my people."

Sullivan tried. Because of his insistence on detailed preplanning of extraction and fire support contingencies, he was a well-known figure at the Direct Air Support Center of the 1st Marine Division. When he called at midnight, he did not bandy words. He wanted flare ships, helicopters and fixed-wing aircraft dispatched immediately to Nui Vu.

Somehow the response was delayed, and shortly after midnight the enemy forces gathered and rushed forward in strength a second time. The Marines threw the last of their grenades and fired their rifles semi-automatically, relying on accuracy to suppress volume. It did, and the enemy fell back - but by that time every Marine had been wounded.

The living took the ammunition of the dead and lay under a moonless sky, wondering about the next assault. Although he did not tell anyone, Howard doubted they could repel a massed charge by a determined enemy. From combat experience he knew that the enemy, having been badly mauled twice, would listen for sounds which would indicate his force had been shattered or demoralized before surging forward again. Already up the slopes were floating the high, singsong taunts Marines had heard at other places in other wars. Voices which screeched, "Marines - you die tonight!" and "Marines, you die in an hour!"

Members of the platoon wanted to return the compliments. "Sure," said Howard, "go ahead and yell anything you want." So the Marines shouted back down the slopes all the curses and invectives they could remember from their collective repertoire. The NVA screamed back, giving Howard the opportunity to deliver a master stroke in psychological one-upmanship.

"All right," he shouted. "Ready? Now!"

All the Marines laughed and laughed and laughed at the enemy.

The North Vietnamese did not mount a third major attack, and at 0100 an Air Force flare ship, with the poetic call sign of "Smoky Gold," came on station overhead. Howard talked to the pilot through his radio and the plane dropped its first flare. The mountainside was lit up, and the Marines looked down the slopes. Lance Corporal Ralph Victor stared, then muttered, "Oh my God, look at them." The others weren't sure it wasn't a prayer. North Vietnamese reinforcements filled the valley. PFC Joseph Kosoglow described it vividly. "There were so many, it was just like an anthill ripped apart. They were all over the place."

They shouldn't have been. Circling above the mountain were attack jets and armed helicopters. With growing frustration, they had talked to Howard but could not dive to the attack without light. Now they had light.

They swarmed in. The jets first concentrated on the valley floor and the approaches to Nui Vu, loosing rockets which hissed down and blanketed large areas. Then those fast, dangerous helicopters - the Hueys - scoured the slopes. At altitudes as low as twenty feet they skimmed over the brush, firing their machineguns in long, sweeping bursts. The Hueys then pulled off to spot for the jets, and again the planes dipped down, releasing bombs and napalm. Then the Hueys scurried back to pick off stragglers, survey the damage, and direct another run. One of the platoon's communicators, Corporal Martinez, said it in two sentences. "The Hueys were all over the place. The jets blocked the Viet Cong off."

Two Hueys stayed over Howard's position all night - when one helicopter had to return to home base and refuel, another would be sent out. The Huey pilots also performed dual roles - they were the Tactical Air Controllers Airborne (TACAs) who directed the bomb runs of the jets, and they themselves strafed the enemy. The NVA tried to shoot the helicopters down, and did hit two out of the four Hueys alternating on station.

By the light of the flares the jet pilots could see the hill mass and distinguish prominent terrain features, but could not spot Howard's perimeter. To mark specific targets for the jets, the TACAs directed "Smoky" to drop flares right on the ground as signal lights and then called the jets down to pulverize the spot. Howard identified his position by flicking a refiltered flashlight on and off, and the Huey pilots strafed to within twenty-five meters of the Marines by guiding on that mark.

Still on the perimeter itself the fight continued. In the shifting light of the flares, the pilots were fearful of hitting the Marines and had to leave some space unexposed to fire in front of the Marines' lines. Into this space crawled the North Vietnamese.

For the Marines it was a war of hide and seek. Having run out of hand grenades, they had to rely on cunning and marksmanship to beat the attackers. Howard had passed the word to fire only at an identified target - and then only one shot at a time. The enemy fired all automatic weapons, and the Marines replied with single shots. The enemy hurled grenades, and the Marines threw back rocks.

It was a good tactic. A Marine would hear a noise and toss a rock in that general direction. The North Vietnamese would think it was a grenade hitting the ground and dive for another position. The Marine would roll or low crawl to a spot from which he could sight in on the position, and wait. In a few seconds, the enemy soldier would raise his head to see why the grenade had not exploded. The Marine would fire one round. The range was generally less than thirty feet.

The accuracy of this fire saved the life of Corpsman Holmes. When he regained consciousness after the grenade had knocked him out, he saw a North Vietnamese dragging away the dead Marine beside him. Then another enemy reached over and grabbed him by the cartridge belt. The soldier tugged at him.

Lance Corporal Victor was lying on his stomach behind a rock. He had been hit twice by grenades since the first flare had gone off and could scarcely move. He saw an enemy soldier bending over a fallen Marine, sighted in, and fired. The man fell backward. He saw a second enemy tugging at another Marine's body. He sighted in and fired again.

Shot between the eyes, the North Vietnamese slumped dead across Billie Holmes' chest. He pushed the body away and crawled back to the Marines' lines. His left arm was lanced with shrapnel, and his face was swollen and his head ringing from the concussion of the grenade. For the rest of the night he crawled from position to position, bandaging and encouraging the wounded - and in between, he fired at the enemy.

Occasionally the flares would flicker out and the planes would have to break off contact to avoid crashing. In those instances, artillery under the control of the Special Forces and manned by Vietnamese gun crews would fill in the gap and punish any enemy force gathering at the base of Nui Vu.

"Stiff Balls," Howard had radioed the Special Forces camp at Hoi An, three miles south. "If you can keep Charlie from sending another company up here, I'll keep these guys out of my position."

"Roger, Carnival Time," Captain Louis Maris of the Army Special Forces replied, using Howard's own peculiar call sign. Both sides kept their part of the bargain, and the Vietnamese crews who manned the 105mm howitzers threw in concentration after concentration of accurate artillery shells.

Howard was talking on the radio. "He was cool," Captain John Blair, the Special Forces commanding officer, recalled afterward. "He stayed calm all the way through that night. But," he chuckled, "he never did get our call sign right!"

During the periods of darkness, each Marine fought alone. How some of them died, nobody knows. But the relieving force hours later found one Marine lying propped up against a rock. In front of him lay a dead enemy soldier. The muzzles of their weapons were touching each other's chests. Two Marine entrenching tools were recovered near a group of mangled North Vietnamese, and both shovels were covered with blood. One Marine was crumpled beneath a dead enemy. Beside him lay another Vietnamese. The Marine was bandaged around the chest and head. His hand still clasped the hilt of a knife buried in the back of the soldier on top of him.

At 0300 a flight of H34 helicopters whirled over Nui Vu and came in to extract the platoon. The fire was so intense they were unable to land, and Howard was told he would have to fight on until dawn. Shortly thereafter, a ricochet struck Howard in the back. His voice over the radio faltered and died out. Those listening - the Special Forces personnel, the pilots, the high-ranking officers of the 1st Marine Division at Chulai - all thought the end had come. Then Howard's voice came back strong. Fearing the drowsing effects morphine can have, he refused to let Holmes administer the drug to ease the pain. Unable to use his legs, he pulled himself from hole to hole, encouraging his men and directing their fire. Wherever he went, he dragged their lifeline - the radio.

Binns, the man whose shot had triggered the battle, was doing likewise. Despite severe wounds he crawled around the perimeter, urging his men to conserve their ammunition, gathering enemy weapons and grenades for the Marines' use, giving assistance wherever needed.

None of the Marines kept track of the time. "I'll tell you this," said Howard, "you know that movie - *The Longest Day*? Well, compared to our night on the hill, *The Longest Day* was just a twinkle in the eye." But the longest night did pass, and dawn came. Howard heralded its arrival. At 0525 he shouted, "Okay you people, reveille goes in thirty-five minutes." At exactly 0600, his voice pealed out, "Reveille, reveille!" It was the start of another day and the perimeter had held.

On all sides of their position, the Marines saw enemy bodies and equipment. The NVA would normally have raked the battlefield clean, but so deadly was the Marine fire that they left unclaimed many of those who fell close to the perimeter.

The firing had slacked off. Although badly mauled themselves, the enemy still had the Marines ringed in and had no intention of leaving. Nor did haste make them foolhardy. They knew what the jets and the Hueys and the artillery and the Marine sharpshooting would do to them on the bare slopes in daylight. They slipped into holes and waited, intending to attack with more troops the next night.

Bursts of fire from light machineguns chipped the rocks above the Marines' heads. Firing uphill from concealed foxholes, the enemy could cut down any Marine who raised up and silhouetted himself against the skyline. Two of the .50-caliber machineguns were still firing sporadically.

There came a lull in the firing. A Huey buzzed low over the hillcrest, while another gunship hovered to one side, ready to pounce if the enemy took the bait. No one fired. The pilot, Major William Goodsell, decided to mark the position for a medical evacuation by helicopter. His Huey fluttered slowly down and hovered. Howard thought the maneuver too risky, and said so. But Goodsell had run the risk and come in anyway. He dropped a smoke grenade. Still no fire. He waved to the relieved Howard and skimmed north over the forward slope, only ten feet above the ground.

The noise of machineguns drowned out the sound of the helicopter's engines. Tracers flew toward the Huey from all directions. The helicopter rocked and veered sharply to the right and zig-zagged down the mountain. The co-pilot, First Lieutenant Stephen Butler, grabbed the stick and brought the crippled helicopter under control, crash landing in a rice paddy several miles to the east. The pilots were picked up by their wingman, but Major Goodsell, who had commanded squadron VMO-6 for less than one week, died of gunshot wounds before they reached the hospital.

The medical pickup helicopter did not hesitate. It came in. Frantically, Howard waved it off. He was not going to see another shot down. The pilots were dauntless, but not invulnerable. The pilot saw Howard's signal and turned off, bullets clanging off the armor plating of the undercarriage. Howard would wait for the infantry.

In anger, the jets and the Hueys now attacked the enemy positions anew. Flying lower and lower, they crisscrossed the slopes, searching for the machinegun emplacements, offering themselves as targets, daring the enemy to shoot.

The enemy did. Another Huey was hit and crashed, its crew chief killed. The .50-calibers exposed their positions and were silenced. Still, the North Vietnamese held their ground. Perhaps the assault company, with all its automatic weapons and fresh young troops, had been ordered to wipe out the few Marines at any cost. Perhaps their commanding officer had been killed and his subordinates were following dead orders. Or perhaps the enemy thought victory was yet possible.

But then the Marine infantry came in. They had flown out at dawn, but the enemy fire around Nui Vu was so intense the helicopters had to circle for forty-five minutes while jets and artillery blasted a secure landing zone. During that time First Lieutenant Richard Moser, an H-34 helicopter pilot, monitored Howard's frequency and later reported, "It was like something you'd read in a novel. His call sign was Carnival Time and he kept talking about these North Vietnamese down in holes in front of him. He'd say, 'You've gotta get this guy in the crater because he's hurting my boys.' He was really impressive. His whole concern was for his men."

On the southern slope of the mountain, helicopters finally dropped Charlie Company of the 5th Marines. The relief company climbed fast, ignoring sniper fire and wiping out small pockets of resistance. With the very first round they fired, the Marines 60mm mortar team knocked out the enemy mortar. Sergeant Frank Riojas, the weapons platoon commander, cut down a sniper at five hundred yards with a tracer from his M-14. Marine machinegun sections were detached from the main body and sent up the steep fingers along the flanks of the hill to support by fire the company's movement. The NVA were now the hunted, as Marines scrambled around as well as up the slope, attempting to pinch off the enemy before they could flee.

The main column climbed straight upward. While still a quarter of a mile away, the point man saw recon's position on the plateau. The boulder which served as Howard's command post was the most prominent terrain feature on the peak. The platoon hurried forward. They had to step over enemy bodies to enter the perimeter. Howard's men had eight rounds of ammunition left.

"Get down," were Howard's first words of welcome. "There are snipers right in front of us." Another recon man shouted, "Hey, you got any cigarettes?" A cry went up along the line - not expressions of joy - but requests for cigarettes.

It was not that Howard's Marines were not glad to see the infantry. It was just that they had expected them. Staff Sergeant Richard Sullivan, who was with the first platoon to reach the recon Marines, said later, "One man told me he never expected to see the sunrise. But once it did, he knew we'd be coming."

The fight was not over. Before noon, in the hot daylight, despite artillery and planes firing in support, four more Marines would die.

At Howard's urging, Second Lieutenant Ronald Meyer quickly deployed his platoon along the crest. Meyer had graduated from the Naval Academy the previous June, and intended to make the Marine Corps his career. He had spent a month with his bride before leaving for Vietnam. In the field he wore no shiny bars, and officers and men alike called him "Stump" because of his short, muscular physique.

Howard had assumed he was a corporal or a sergeant and was shouting orders to him. Respecting Howard's knowledge and performance, Meyer obeyed. He never did mention his rank. So Staff Sergeant Howard, waving off offers of aid, proceeded to direct the tactical maneuvers of the relieving company, determined to wipe out the small enemy band dug in not twenty meters downslope.

Meyer hollered for members of his platoon to pass him grenades. He would then lob them downslope toward the snipers' holes. By peering around the base of the boulder,

Howard was able to direct his throws. "A little more to the right on the next one, buddy. About five yards further. That's right. No, a little too strong." The grenades had little effect, and the snipers kept firing. Meyer shouted he wanted air on the target. The word was passed back for the Air Liaison Officer to come forward. The platoon waited.

Lance Corporal Terry Redic wanted to fire his rifle grenade at the snipers. A tested sharpshooter, he had several kills to his credit. In small fire fights he often disdained to duck, preferring to suppress hostile fire by his own rapid, accurate shooting. Meyer's way seemed too slow. He raised up, knelt on one knee, and sighted downslope looking for a target. He never found one. The enemy shot first, and killed him instantly.

Meyer swore vehemently. "Let's get that SOB! You coming with me, Sotello?"

"Yes, Stump."

Lance Corporal David Sotello turned to get his rifle and some other men. Meyer didn't wait. He started forward with a grenade in each hand. "Keep your head down buddy, they can shoot," yelled Howard.

Meyer crawled for several yards, and then threw a grenade at a hole. It blasted an enemy soldier. He turned, looking upslope. Another sniper shot him in the back. Sotello heard the shot as he started to crawl down.

So did Hospital Corpsman 3rd Class John Markillie. He crawled toward the fallen lieutenant. "For God's sake, keep your head down!" yelled Howard. Markillie reached his lieutenant, and sat up to examine the wound. A sniper shot him in the chest.

Another corpsman by the name of Holloday and a squad leader, Corporal Melville, crawled forward. They could not feel Meyer's pulse. Markillie was still breathing. Ignoring the sniper fire, they began dragging and pushing his body up the hill.

Melville was hit in the head, and he rolled over as his helmet bounced off. He shook his head and continued to crawl. The round had gone in one side of his helmet and ripped out the other, just nicking the corporal above his left ear. Melville and Holloday finally dragged Markillie into the perimeter.

From Chulai, the battalion commander called his company commander, First Lieutenant Marshall "Buck" Darling. "Is the landing zone secure, Buck?"

"Well," a pause..., "not spectacularly."

Back at the base two noncommissioned officers were listening. "I wonder what he meant by that?" asked the junior sergeant.

"What the hell do you think it means, stupid?" replied the older sergeant. "It means he's getting shot at."

Ignoring his own wounds, Corpsman Billie Holmes was busy supervising the corpsmen from Charlie Company as they administered to the wounded. With the fire fight still going on to the front, helicopter evacuation was not possible from within the perimeter. The wounded had to be taken rearward to the south slope. Holmes roved back and forth, making sure that all his buddies were accounted for and taken out.

The pilots had seen easier landing sites. "For the medical evacs," Moser said, "a pilot had to come in perpendicular to the ridge, then cock his bird around before he sat down. We could get both main mounts down. The tail, well, sometimes we got it down. We were still taking fire."

Holmes reported that there was still one Marine, whom he had seen die, missing. Only after repeated assurances that they would not leave without the body were the infantry able to convince him and Howard that it was time they too left. They helped the Navy corpsman and the Marine sergeant to a waiting helicopter. Howard's job was done.

Another had yet to be finished. There was a dead Marine to be found somewhere on the field of battle, but before a search could be conducted the last of the enemy force had to be destroyed.

First Lieutenant Phil Freed flopped down beside Melville. Freed was the Forward Air Controller attached to Charlie Company that day, and had run the last quarter mile uphill when he heard Meyer needed air. With rounds cracking near his head, he needed no briefing. He contacted two F8 Crusader jets circling overhead. "This is Cottage 1-4. Bring it on down on a dry run. This has to be real tight. Charley is dug in right on our lines." At the controls of the jets were First Lieutenants Richard Deilke and Edward Menzer.

"There were an awful lot of planes in the air," Menzer said. "We didn't think we'd be used, so we called the DASC (Direct Air Support Center) and asked for another mission. We got diverted to the FAC, Cottage 1-4. He told us he had a machinegun nest right in front of him."

As they talked back and forth Menzer thought he recognized Freed's voice. Later he learned he had indeed - Freed had flown jets with him in another squadron a year earlier.

Freed was lying in a pile of rocks on the military crest of the northern finger of the hill. Since he himself had flown the F8 Crusader, he could talk to the pilots in a language they understood. Still, he was not certain they could help. He didn't know whether they could come that close and still not hit the Marines on the ground. On their first run he deliberately called the jets in wide so he could judge the technical skills and precision of the pilots. Rock steady.

231

Now he called for them to attack in earnest. When they heard the target was twenty meters from the FAC, it was the pilots' turn to be worried. "As long as you're flying parallel to the people, it's OK," Menzer said, "because it's a good shooting bird. But even so, I was leery at first to fire with troops that near."

Unbeknownst to them, the two pilots were about to fly one of the closest direct air support missions in the history of fixed-wing aviation. They approached from the northeast with the sun behind them, and cut across the ridgeline parallel to the friendly lines. They strafed without room for error. The gunsight reflector plate in an F8 Crusader looks like a bull's-eye with the rings marked in successive 10-mil increments. When the pilots in turn aligned their sights while three thousand feet away, the target lay within the 10-mil ring and the Marine position was at the edge of the ring. The slightest variance of the controls would rake the Marine infantrymen with fire. In that fashion each pilot made three strafing passes, skimming ten to twenty feet above the ridge. Freed feared they would both crash, so close did their wings dip to the crest of the hill. The impact of the cannon shells showered the infantrymen with dirt. They swore they could tell the color of the pilot's eyes. In eight attacks the jets fired 350 20mm explosive shells into an area sixty meters long and ten to twenty meters wide. The hillside was gouged and torn, as if a bulldozer had churned back and forth across it.

Freed cautiously lifted his head. A round cracked by. One enemy had survived. Someone shouted that the shot came from the position of the sniper who had killed Meyer. The lieutenant's body lay several yards down the slope.

The F-8 Crusaders had ample fuel left. Menzer called to say they could make dummy runs over the position if the

Marines thought it would be useful. Freed asked them to go ahead and try it.

The company commander, Buck Darling, watched the jets. He noticed that whenever they passed over the firing stopped momentarily. The planes would be his cover. "I'm going to get Stump. Coming, Brown?" he asked the nearest Marine.

Lance Corporal James Brown was not a poster Marine. His offbeat sense of humor often conflicted with his superiors' sense of duty. His squad leader later recalled with a grimace one fire fight when the enemy caught the squad in a cross fire. The rounds were passing high over the Marines' heads. While everyone else was returning fire Brown strolled over to a Vietnamese tombstone, propped himself against it with one finger, crossed his legs and yelled, "You couldn't hit me if I was buried here!" His squad leader almost did the job for the enemy.

On the hill relieving the recon unit, however, Brown was all business. He emptied several rifle magazines and hurled grenade after grenade. When he ran out of grenades, he threw rocks to keep the snipers ducking. All the while he screamed and cursed, shouting every insult and blasphemy he could think of. Howard had been very impressed, both with Brown's actions and with his vocabulary.

He was not out of words when Darling asked him to go after Meyer's body. As they crawled over the crest, Brown tugged at his company commander's boot. "Don't sweat it, Lieutenant, they can only kill us." Darling did not reply. They reached Meyer's body and tried to pull it back while crawling on their stomachs. They lacked the strength.

"All right, let's carry him," said Darling. It was Brown's turn to be speechless. He knew what had happened to every Marine on the slope who had raised his head - and here was his officer suggesting they stand straight up! "We'll time our moves with the jets." When the jets passed low, they stumbled and scrambled forward a few yards with their burden, then flattened out as the jets pulled up. The sniper snapped shots at them after every pass. Bullets chipped the rocks around them. They had less than thirty feet to climb. It took over a dozen rushes. When they finally rolled over the crest they were exhausted. Only the enemy was left on the slope.

The infantry went after him. Corporal Samuel Roth led his eight man squad around the left side of the slope. On the right, Sergeant Riojas set up a machinegun on the crest to cover the squad. A burst of automatic fire struck the tripod of the machinegun. A strange duel developed. The sniper would fire at the machinegun. His low position enabled him to aim in exactly on the gun. The Marines would duck until he fired, then reach up and loose a burst downhill, forcing the sniper to duck.

With all the firing, the sniper could not hear the squad crashing through the brush on his right side. Roth brought his men on line facing the sniper. With fixed bayonets they began walking forward. They could see no movement in the clumps of grass and torn earth.

There was a lull in the firing. The sniper heard the squad, turned, and fired. Bullets whipped by the Marines. Roth's helmet spun off. He fell. The other Marines flopped to the ground. Roth was uninjured. A steel helmet had saved a second Marine's life within an hour. He was not even aware that his helmet had been shot off. "When I give the word, kneel and fire," he said. "Now!"

Just as the Marines rose, the sniper bobbed up like a duck in a shooting gallery. A bullet knocked him backwards against the side of his hole. Roth charged, the other Marines sprinting behind him. He drove forward with his bayonet. A grenade with the pin intact rolled from the sniper's left hand. Roth jerked the blade back. The sniper slumped forward over his machinegun.

The hill was quiet. It was noon. Darling declared the objective secure. In the tall grass in front of Riojas' machinegun, the missing Marine was found. The Marines paused to search thirty-nine enemy dead for documents, picked up eighteen automatic weapons, climbed on board a flight of helicopters, and flew off the mountain.

The Marines had lost a total of ten dead. Charlie Company and the Huey squadrons each lost two. Of the eighteen men in the reconnaissance platoon six were killed, and the other twelve wounded. Five members of Charlie Company were recommended for medals. Every Marine under Howard's command received the Purple Heart. Thirteen received the Silver Star. Binns, Holmes, Gerald Thompson and John Adams got the Navy Cross. And Howard was awarded the Medal of Honor.

If the incident had centered around just one man, then it could be considered a unique incident of exceptional bravery on the part of an exceptional man. It was that. But perhaps it was something more. On 14 June, few would have noticed anything unique about the 1st Reconnaissance Platoon of Charlie Company. Just in reading the names of its dead, one has the feeling that here are the typical and the average who, well trained and well led, rose above normal expectations to perform an exemplary feat of arms: John Adams, Ignatius Carlisi, Thomas Glawe, James McKinney, Alcadio Mascarenas, and Jerrald Thompson.

THE WHITE FEATHER

Gunnery Sergeant Carlos N. Hathcock II

"There is no hunting like the hunting of man, and those who have hunted armed men long enough and liked it, never care for anything else thereafter." – Ernest Hemingway

On May 20th, 1959, at seventeen years of age, Carlos N. Hathcock II fulfilled his childhood dream by enlisting in the United States Marine Corps. His exceptional ability as a marksman was soon recognized by the instructors on the rifle range at Camp Pendleton, where he was undergoing recruit training. Later, while based in Hawaii as a member of Company E, 2nd Battalion, 4th Marine Regiment, Hathcock won the Pacific Division rifle championship. Following his assignment in Hawaii, Carlos was transferred to Marine Corps Air Station, Cherry Point, North Carolina where he quickly found himself shooting competitively again. This time he set the Marine Corps record on the "A" Course with a score of 248 points out of a possible 250, a record that stands today. The highlight of his competitive shooting career occurred in 1965 when Carlos out-shot over three thousand other servicemen competing to win the coveted Wimbledon Cup at Camp Perry.

This achievement led to his being sought out in Vietnam in 1966 to be part of a newly established sniper program. Before being assigned as a sniper, Hathcock had served as a military policeman, and even then he demonstrated a natural gift for sniping:

"A flash of movement caught the eye of a young Marine Military Policeman who was keeping watch for possible enemy action. As he observed, he could make out a figure crouched in the distance, working busily with something he couldn't quite see. The man was in civilian clothes... but... there was the rifle slung over his back - the telltale mark of a Viet Cong guerrilla. The enemy soldier continued about his task, oblivious to his danger as Sergeant Carlos Hathcock brought his M-14 to bear. The range appeared to be between 300 and 400 yards – child's play for Hathcock, who had won the 1000 yard Wimbledon Cup Match at Camp Perry only the year before. The rack grade weapon he now held was a far cry from the finely-fitted National Match M-1 he had used in competition, but it was certainly capable of making this shot. With his M-14 rested comfortably, Hathcock verified his target - yes, definitely armed - and adjusted his position slightly. He let the front sight settle naturally, centered on the crouching soldier, who appeared to be placing a booby trap.

Hathcock felt his chest tighten and his heartbeat increase. Although already a Distinguished and a world-class competitive rifleman, he was still new to combat and the killing of men. As he silently eased the safety forward, his right hand settled firmly into place on the small of the stock. He was in his "bubble" now - a zone of total concentration. He exhaled, and there was the front sight: on target, crisp, in razor-sharp focus and centered in the rear sight aperture. The rifle was absolutely still as he took up the slack in the two-stage trigger, and then applied the final pressure. Such was the depth of his concentration that he was only vaguely aware of the rifle's report as it jolted against his shoulder. As the bolt cycled, the empty case skittered brightly across the ground to his right, and the M-14 settled back into position, cocked and ready for a second shot. None was needed, however. The enemy guerrilla lay sprawled, no longer a threat. Sergeant Carlos Hathcock had made his first kill. Officially, it was unconfirmed - one of fourteen unconfirmed kills he was to make before his assignment as a Marine sniper. However, that didn't concern him. It was simply a job that had to be done. By his actions, Carlos Hathcock had certainly saved the lives of several brother Marines scheduled to patrol the area being mined that day."

It wasn't long before Hathcock underwent formal instruction as a sniper, and after his training was complete he began his new assignment. While operating from Hill 55, a position thirty-five miles Southwest of Da Nang, Hathcock and his fellow Marine snipers renewed a Marine tactic which had been born in the islands of the Pacific during World War II. Within a short period of time the effects of the Marine snipers could be felt around Hill 55, and Carlos rapidly ran up a toll on the enemy that would eventually lead to a $30,000 bounty being placed on his head by the NVA.

As a result of his skill Sergeant Hathcock was twice recruited for covert assignments. One of them was to kill a Frenchman who was working for the North Vietnamese as an interrogator. This individual was torturing American airmen who had been shot down and captured, but one round from the Marine sniper's modified Winchester Model 70 ended the Frenchman's career. On another occasion Sergeant Hathcock accepted an assignment for which he was plainly told his odds for survival were slim. A North Vietnamese general was the target, and the man died when a bullet fired by Carlos struck him from a range of eight hundred yards. In order to make the shot, Hathcock had to cover more than one thousand meters of open terrain during three days and nights of constant crawling, an inch at a time. Enemy patrols came within twenty feet of Hathcock, who lay camouflaged with grass and vegetation in the open. The Marine sniper returned to Hill 55 unscathed.

In another incredible incident, an enemy sniper was killed after a prolonged game of "cat and mouse" between Carlos, with his spotter, and the NVA sniper. The fatal round, fired at five hundred yards by Hathcock, passed directly through the NVA sniper's rifle scope, striking him in the eye.

Hathcock would eventually be credited with ninety-three enemy confirmed killed, including one Viet Cong shot dead by a round fired from a scope-mounted Browning M-2 .50-caliber machine gun at the unbelievable range of 2500 yards. By this time the Viet Cong knew him well, and called him "Long Trang" - Vietnamese for "white feather" - because he often wore one in his bush hat to taunt them.

Hathcock's career as a sniper came to a sudden end outside Queson in 1969, when the amphibious tractor he was riding in was ambushed and hit a 500-pound box mine. Hathcock pulled seven Marines out of the flame-engulfed vehicle before jumping to safety himself. He came out of the attack with second and third-degree burns over more than forty percent of his body, and was evacuated to Brooke

Army Medical Center in Texas, where he underwent thirteen skin graft operations. Sadly, the nature of the injuries left him unable to perform effectively again with a rifle.

Despite the severity of his wounds, it would ultimately be the ravages of Multiple Sclerosis (MS) that would bring Hathcock's extraordinary career to an end, and in 1979 he was forced to retire on 100% disability due to the advancing stages of the disease.

Gunnery Sergeant Hathcock spent subsequent years instructing police tactical units in "counter-sniper" techniques, and in 1990 a book entitled *Marine Sniper* was published which documented the exploits of this one-of-a-kind Marine.

The Marine Corps eventually named the annual Carlos Hathcock Award after him, which is given to its best marksman. A Marine library in Washington has also been named for him, and a Virginia Civil Air Patrol unit took his name as well. In 1990 a Marine unit raised $5,000 in donations to fight multiple sclerosis, and presented it to him at his home - and they brought it to him the old-fashioned, Marine Corps way, running the 216 miles from Camp Lejeune, North Carolina, to Virginia Beach.

During his retirement ceremony, Hathcock was presented a plaque by his commanding officer. It read:

"There have been many Marines. And there have been many Marine marksmen. But there is only one Marine Sniper - Gunnery Sergeant Carlos N. Hathcock II. One Shot - One Kill."

Gunnery Sergeant Carlos Hathcock died on February 26, 1999

DEWEY CANYON

First Lieutenant Wesley Fox

"The warrior has come center stage once again, and it is time for the tender-hearted to take a seat and be quiet. The warrior will ensure that they, the talkers, retain all of their rights, to include letting the warrior do the tough, ugly work." - Colonel Wesley Fox

Running almost the entire length of both North and South Vietnam was an enemy supply road, infamously known as the Ho Chi Minh trail. Stretching from the Communist capitol in the north, this supply route funneled thousands of enemy soldiers and tons of weapons and supplies to the south, almost entirely within the protected confines of Laos and Cambodia. But for aerial attack, the Ho Chi Minh trail was

243

invincible. United States soldiers and Marines were not allowed to cross the borders.

During the Christmas cease-fire of 1967 the North Vietnamese used the Ho Chi Minh trail to amass tens of thousands of soldiers for a major offensive. A month after Christmas, on January 30, 1968, nearly 100,000 enemy soldiers launched a major offensive, striking simultaneously at every provincial capitol in the south. In the northwest corner of South Vietnam, embattled Marines survived a 77-day siege at Khe Sahn. Throughout the northern portion of South Vietnam, labeled by the military as I CORPS, fierce fighting raged for months. The massive enemy buildup had been staged from within their Laotian sanctuary, and launched in I Corps in large part from fortified positions deep inside a South Vietnamese mountainous jungle called the A Shau Valley.

The A Shau Valley was one of two major enemy strongholds in the south. The other was the U Minh Forest. Twenty-two miles long, the A Shau Valley was only six miles from the Laotian border, a deep valley that ran between two heavily forested mountain ranges. The strongest enemy base in South Vietnam, A Shau was protected by a sophisticated complex of interlocked anti-aircraft batteries and garrisoned by more than five thousand enemy soldiers.

Following the Tet Offensive of 1968, signs of American presence in A Shau were limited primarily to three abandoned airfields spread throughout the valley floor, and a deserted Special Forces camp at the southern tip of the valley. The Special Forces camp had been overrun by the enemy in 1966. For the most part, the A Shau was a staging point where the enemy could build huge stockpiles of weapons and supplies funneled south on the trail through Laos, and then launch strikes against American and South

Vietnamese troops throughout the Quang Tri and Thua Thien Provinces.

A year after the Tet Offensive, military intelligence reports indicted a massive enemy buildup in the already heavily enemy-controlled A Shau. Plans at this stage of the war were for a decreasing role for U.S. ground troops, and transfer of responsibility for combat actions to the soldiers of the Army of South Vietnam (ARVN). But the enemy strength in A Shau posed a threat that demanded an immediate American effort to deny the enemy his sanctuary, capture his supplies, and prove that the A Shau would no longer be a haven. Primary responsibility for this mission fell to the men of the 3rd Marine Division's 9th Marine Regiment.

Headquartered fifty miles northeast of the A Shau at Vandergrift Combat Base under the command of Colonel Robert H. Barrow, the 9th Marines boasted three battalions to be marshaled for this formidable task. The mission would be one of the last major offensives conducted by U.S. Marines in Vietnam. It would be tough. It would be deadly. But it would be, in the tradition of the U.S. Marine Corps, an engagement fought valiantly and successfully.

Lance Corporal Thomas Noonan did his best to ignore the mud as his company slowly moved down the side of the hillside. It was February 5th, and Operation Dewey Canyon was two weeks old. The men of the 9th Marines were moving into the A Shau Valley, and Company G, 2nd Battalion, 9th Marines was moving out of its location southeast of Vandergrift Combat Base as part of the incursion into A Shau. It was early in the monsoon season, so the progress of the Marines was hampered not only by the dense foliage, but also by the intermittent rains and the slippery mud. Suddenly, bad went to worse as the lead element walked into the enemy.

The North Vietnamese opened fire from their concealed positions, wounding four men. The rest of the Marines were stuck further up the hill by the impossible terrain and the hail of enemy fire. No one could reach the wounded.

Lance Corporal Noonan was a rifleman in his company, despite an education that could have gotten him an almost immediate commission when he joined the Corps. Now he took upon himself the task of rescuing his brothers, and carefully he moved down the slippery slope - ever mindful of the heavy enemy presence. Nearing the wounded, he took cover behind some rocks and shouted encouraging words to the wounded Marines, assuring them help was on the way.

He then raced alone across the fire-swept area, located the most seriously wounded man and dragged him backwards toward shelter. Enemy rounds whistled through the area, hitting Noonan and knocking him to the ground. Despite his own wounds, Lance Corporal Noonan got back up and resumed dragging the wounded Marine towards the cover of the rocks from which he had yelled encouragement. Before he reached its shelter enemy fire reached out again, a rain of hot lead striking the young corporal's body. Inspired by Noonan's example, the rest of the platoon suddenly charged the enemy and pushed them back, allowing them to reach the wounded. All four survived, with Thomas Noonan being the only casualty. He died with the collar of his wounded comrade's fatigue shirt still grasped in his hands, and would be the first to earn a Medal of Honor in an operation that would test the courage of every man in the 9th Marine Regiment.

The commanding general of the 3rd Marine Division, headquartered out of Da Nang, was Major General Raymond G. Davis. This was Davis' third war, having served in World War II and received the Medal of Honor for his heroism at the Chosin Reservoir in Korea. Davis referred to the 9th Marines as the "Mountain Regiment" and his "Strike Force Regiment." As Operation Dewey Canyon began on January 22nd General Davis had good reason to pay close attention to the efforts of his Marines, because among the men assigned to meet and defeat the enemy in their A Shau sanctuary was a young lieutenant in command of a rifle platoon. By a strange twist of fate which defied military policy prohibiting relatives from serving in the same war zone, Operation Dewey Canyon would send Davis' son, Lieutenant Miles Davis, into harm's way.

The first phase of Operation Dewey Canyon primarily involved the movement and positioning of air assets. Phase II, the movement of the three battalions of the 9th Marines out of Vandergrift Combat Base, began on January 31st. It was during this effort to position the 9th Marines at the northern edge of the A Shau that Lance Corporal Noonan died trying to save the life of his fellow Marines.

From January 31st until February 10th 2/9 continued its movement south, flanked by 1/9 and 3/9. Colonel Robert Barrow, commander of the 9th Marine Regiment (who later became the 27th Commandant of the Marine Corps), coordinated the mission with support from American assets throughout I Corps. By February 10th the three battalions were poised and ready to enter Phase III, the incursion into A Shau. Along the way they had built numerous fire bases with names like Henderson, Tun Tavern, Shiloh, Razor, and Cunningham to provide artillery support and maintain supply routes.

The large movements of the three battalions demanded a regular and consistent resupply at Vandergrift Combat Base. On February 13th, far to the north, a convoy was carrying supplies to Vandergrift when it was ambushed. In the heavy mortar and small arms fire that followed, the convoy security squad moved to engage the enemy. When Lance Corporal Thomas Creek dashed across the fire swept area to take up a better position from which to attack the enemy, he was wounded. Almost as quickly as he fell, the enemy threw a grenade at his position. Nearby were others of his squad, men who were about to die. Ignoring his previous wounds, Lance Corporal Creek rolled on top of the grenade to absorb the blast, sparing the lives of his comrades at the expense of his own. Though not officially a part of the Dewey Canyon operation, in his support role the eighteen-year old hero's posthumous Medal of Honor must be counted with that of those who fought further south.

Even as Lance Corporal Creek lay dying in a gully northeast of Vandergrift and far to the south, 3/9 was crossing the Da Krong River only thirteen kilometers from Laos. Phase II of Operation Dewey Canyon was underway. The following morning the Marines of 1/9 and 2/9 began moving out of their fire bases as well, heading southward and towards the North Vietnamese Base Area 611 that ran from the north boundary of A Shau and into Laos.

The move into A Shau was both miserable and dangerous. Triple-canopy jungle made movement difficult, and two weeks of continuous fog and heavy monsoon rains removed any possibility of personal comfort and made resupply difficult. The enemy moved freely through the A Shau at night on roads they would carefully camouflage during the day with movable trees and shrubs ingeniously planted in containers. As they moved, the Marines were subjected to heavy artillery fire from NVA guns inside Laos. At night, more troops and weapons moved down the Ho Chi Minh trail. General Davis caused a slight stir on the home front when a newspaper reported a remark made in a personal conversation. "It makes me sick," the 3rd Marine Division CG had said, "to sit on this hill and watch those thousand (enemy) trucks go down those roads in Laos, hauling ammunition down south to kill Americans with."

Early in the third phase of the operation, General Davis flew out to meet with Colonel Barrow. Earlier that morning the General had begun receiving reports of enemy contact... "Kilo Company, fire fight, one killed, two wounded." An hour later Kilo Company was in contact again, with more Marines killed, and more wounded. Davis did his best to concentrate on the meeting at hand, all the while knowing his son was a rifle platoon leader in Kilo Company. Even as he landed for his meeting with Colonel Barrow, some of the wounded were arriving... and Lieutenant Miles Davis was among them. The younger Davis would survive his wounds, and subsequently would receive the Purple Heart Medal from his father.

Meanwhile in the A Shau, other Marines struggled to stay alive and complete their mission. By February 20th the Marines had moved all the way to the Laotian border. As the enemy played their deadly game of hide-and-seek, raining death on young American Marines and then quickly scurrying across the border into the safety of Laos, Colonel Barrow had seen too many of his men die due to the unfair advantage. "The political implications of going into Laos were pretty unimportant to me at that point," he later stated.

The policy of U.S. commanders had always been that units could enter Laos or Cambodia, but only when American lives were endangered by enemy forces therein. (Usually this applied to SAR (Search and Rescue) missions for downed pilots or LRRP (long range reconnaissance patrols). Colonel Barrow saw the danger his own men faced, and despite the very real possibility of sacrificing his own distinguished military career, ordered Hotel Company, 2/9 to cross the border and set up ambush positions *inside* Laos. (This plan was approved by General Creighton Abrams, commander of all U.S. Forces in Vietnam... *after* the border crossing had already occurred.)

With elements of the 9th Marines now operating inside Laos, the other battalions moved out to take up positions along the border. On the morning of February 22nd, the 1st Battalion was in place on a ridge overlooking Laos. The Marines of 1/9 called themselves the "Walking Dead." On this day, for one company in particular, the name would be all too real.

First Lieutenant Wesley Fox was in command of Alpha Company, 1/9. He was a seasoned veteran, now in his nineteenth year in the Marine Corps. He had served in Korea as a young corporal, slowly working his way through the ranks to become a First Sergeant by 1966. The veteran leatherneck, now in his second war, had begun anew... working his way through the commissioned ranks. Fox had already completed a tour in Vietnam, and had recently extended his combat tour.

As dawn broke on the forested hillside overlooking Laos, Alpha Company was sent out to look for and destroy a suspected enemy force operating in the region. Lieutenant Fox's 3rd platoon had made contact with them the previous day, and now the Company was looking to finish the fight. In addition, First Battalion was low on water. A detail was dispatched to get resupply from a stream below, with Alpha Company leading the way to provide security as Lieutenant Fox and his men searched for the enemy. As they reached the stream, the enemy appeared. The NVA seemed to be everywhere, popping up out of hidden spider holes to rain devastating machinegun and small arms fire on Alpha

Company, while enemy mortars fell on the embattled Marines. The suddenness and the ferocity of the attack caught the Marines by surprise, with many falling wounded in the initial onslaught.

Quickly Lieutenant Fox moved out, working his way through the heavy jungle overgrowth to gain a position where he could assess the situation and direct his platoon leaders. Deadly missiles struck the foliage and bamboo palms around him. Fox located a sniper's position, and quickly killed the enemy with his M-16 rifle before moving on.

As Fox deployed his platoons two enemy mortar rounds landed in his position, killing his radiomen and the air and artillery observers. Shrapnel stuck the lieutenant in the shoulder, but despite the bleeding wounds he grabbed both radios and continued to direct the movements of his Marines.

The lieutenant who led Fox's 2nd platoon was seriously wounded, and Fox instructed his executive officer to take command of that platoon. When his platoon leader in the 3rd platoon was killed, Fox quickly moved in to fill the void and take command. He personally destroyed one position while continuing to shout orders and give encouragement. Coolly he spoke into the radio to coordinate aerial and artillery support for his Marines. Among those working to defend these Marines on the Laotian border was artillery officer Harvey "Barney" Barnum, who had earned the Medal of Honor three years earlier and returned, at his own request, for another Vietnam tour.

As the enemy fire continued unabated, the executive officer Fox had sent to 2nd Platoon was killed, and another of his lieutenants was wounded. Though wounded himself, Fox was the only officer in Alpha Company still capable of leading. This he did with calm professionalism, and his

Marines repulsed a final enemy assault during which the Company Commander was wounded a second time.

Heedless of his battered body, Fox began organizing his survivors and establishing a defensive position. As corpsmen moved about to locate and treat the wounded Fox refused aid, setting himself to the tasks leadership demanded. By late afternoon his Marines had secured their position, and Delta Company 2/9 arrived to relieve them. Ten of Fox's brave Marines had died, and of the 153 men who had joined him that morning in the patrol down from the ridge, only sixty-six were able to continue the mission the following day. Despite his wounds, and determined not to leave Alpha Company leaderless, Lieutenant Fox was among them.

For two weeks Captain Dave Winecoff and the Marines of Hotel Company operated in Laos, setting up ambush positions to strike back at an enemy that had never recognized the neutrality of Laos, and who had effectively used it to advantage against American soldiers and Marines. Five days after the border crossing, one patrol from Hotel 2/9 was moving down a road inside Laos when it was attacked by an NVA squad firing automatic weapons and rocket-propelled grenades from the shelter of their fortified bunkers.

Two Marines fell very close to the enemy bunkers, wounded and trapped. Efforts to reach them by others in their patrol were met by heavy and sustained enemy fire. Twenty-one year old Corporal William Morgan slowly began working his way through the undergrowth to a position near the open road directly in front of the enemy position. Yelling encouragement to the wounded Marines, Morgan suddenly got to his feet and rushed into the open roadway to single-handedly attack the enemy. Fully exposed to the fury of the enemy emplacement, his efforts drew an immediate and torrential rain of enemy fire. Corporal Morgan's lifeless body fell to a heap in the middle of the roadway, but his valiant charge had distracted the enemy and drawn their fire long enough for the remaining men of the patrol to reach and rescue the wounded. A year later Corporal Morgan's family accepted his Medal of Honor from President Richard Nixon during ceremonies at the White House. His date of heroism is listed as 25 February 1969, and the place of action is listed vaguely as "southeast of Vandegrift Combat Base, Quang Tri Province, Republic of Vietnam." Indeed, he sacrificed his life southeast of Vandegrift... far to the southeast... far enough to be outside Quang Tri province and inside the borders of Laos.

Operation Dewey Canyon officially ended on March 3rd 1969 when Hotel Company left Laos. In those fifty-six days, the Marines built more than twenty base camps and accounted for more than two thousand enemy casualties. During the period more than five hundred tons of enemy ammunition was confiscated or destroyed, as well as more than one hundred tons of rice stockpiled to feed the enemy troops. Much of the confiscated rice was subsequently donated to needy villages in the region.

On the last official day of Operation Dewey Canyon a reconnaissance company was returning to their base of operations out of Fire Support Base Cunningham when they were attacked. Private First Class Alfred Mac Wilson took charge as acting squad leader of the rear squad, maneuvering his men to outflank the enemy. When two men manning the machinegun were wounded, Alfred Wilson and another Marine were rushing to man the gun themselves when an enemy soldier stepped from behind a tree to lob a grenade between them. Immediately, Corporal Wilson threw his own body over the grenade, dying beneath its lethal blast, to spare his comrade.

Hostilities did not end with the official conclusion of Dewey Canyon either. The Marines continued to hold their base camps in and around A Shau as the Mountain Regiment began their slow withdrawal. Two days after Alfred Wilson jumped on a grenade to save his comrade, a similar engagement occurred at Fire Support Base Argonne near the DMZ, where Marines of the 3rd Reconnaissance Battalion were lending their own support to the Dewey Canyon Operation.

When the enemy made an early morning attack on a twelve-man recon team's position, Private First Class Fred Ostrom headed for a two-man position with his friend, PFC Robert Jenkins. An enemy grenade exploded as the men reached their position, blowing off part of Ostrom's right hand and arm. As the wounded man sagged into his position, a second grenade was thrown directly into the emplacement. Jenkins quickly pushed Ostrom aside, and then covered his wounded friend with his own body. Jenkins shield of flesh and blood spared the life of his wounded friend, but at the cost of his own life.

Such was the courage of the Marines, from experienced vets of other wars like General Davis, Colonel Barrow and Lieutenant Fox, to the valiant young warriors fresh out of high school. They went, they did their job, and some came home. Valor abounded, and many were the acts of heroism which went unrecognized. In the end five Marines earned Medals of Honor, but only one, Lieutenant Wesley Fox, survived to wear it.

He does so humbly, in honor of all those valiant Marines who became a part of his own legacy during Operation Dewey Canyon.

THE BRIDGE AT DONG HA

Captain John Ripley

"War educates the senses, calls into action the will, perfects the physical constitution, brings men into such swift and close collision in critical moments that man measures man." - Emerson

By the Spring of 1972 the North Vietnamese Army (NVA) had completed its buildup and was ready to mount a large-scale attack on South Vietnam. At midday on March 30th, 1972, almost by complete surprise, the North Vietnamese Army launched its single biggest assault of the Vietnam

War. Larger in size and scale than the very costly but politically effective 1968 Tet Offensive, the NVA this time were fighting an almost conventional battle.

Generously supplied with seemingly unlimited artillery, Soviet armor and the latest air-defense weapons, reports of NVA strength and battlefield successes were, for the first few days, not believed by the South Vietnamese general staff and their senior American advisors in Saigon. As part of the assault two infantry divisions, thirty thousand soldiers with tanks and artillery support, began to cross the boundary between the two countries and attack south along Highway 1, the main north-south artery.

The first three and a half days of what came to be known as the Easter Offensive of 1972 were a near rout. The shock value of the new conventional NVA juggernaut was wreaking havoc on friendly forces. Indiscriminate artillery barrages, as intense as any experienced by the old hands, were especially deadly for the uncounted masses of peasants turned refugees in Quang Tri Province. The poor weather and low visibility temporarily neutered the South's advantage in air power. It was hard to believe things could turn so negative in such a short time.

It was clear early on that the town of Dong Ha was a strategic target for the NVA. Offering the only bridge over the Cam Lo-Cua Viet River capable of supporting the heavy T-54 tanks now being used with such tremendous effect, the enemy needed to take it intact. Control of that one bridge would open the South for further exploitation. At a minimum, the capture of Dong Ha would assure the loss of the northern provinces.

The allied unit closest to the gathering storm at Dong Ha was the Vietnamese Third Marine Battalion. As fate would have it, Captain John Ripley was the covan (the Vietnamese name "co-van" for U.S. Marine Corps advisers means "trusted friend") that day about to enter the arena. Major Le Ba Binh commanded the Third Battalion, and had a record every bit as impressive as his American adviser. Wounded on a dozen occasions and decorated many times, he was noted for leading his men from the front as would be expected from a member of the aristocratic warrior class.

By 1971, John Ripley had done almost everything a Marine captain could accomplish commensurate with his rank. Having already successfully served in Vietnam as an infantry company commander in 1967, during which time Ripley was decorated and wounded, he'd had subsequent tours with Marine Force Recon and as an exchange officer with the British Royal Marines. That is particularly noteworthy, because postings with the Royal Marines are extremely competitive and go only to the most promising officers. Happily married and the father of three very young children, Ripley did not really need to be back in Vietnam - but there he was.

The ferocity of the NVA offensive caused all manner of problems with allied command and control. Due to the extreme emergency, Lieutenant Colonel Gerry Turley, who had recently arrived to serve as the senior covan in the northern region, was ordered to also assume control of the Third ARVN Division Forward. Recognizing the need to destroy the bridge, and even though higher headquarters - who were unaware of the deteriorating tactical situation - ordered him not to, Turley gave the order. He was certain he was sending Captain Ripley to his death.

With some cover fire provided by the men of the Third Marine Battalion and aided by U.S. Army Major John Smock, Captain John Ripley accomplished what was not possible - he went out and blew up the bridge.

There is no sports analogy for what Ripley did. It was not like running a three minute mile, bench pressing seven hundred pounds, or pulling out a come-from-behind Super Bowl upset victory. There were no adoring crowds. What Ripley did was simply impossible. Had he failed while attempting to do it, his peers would have only thought him noble and brave for trying.

The significance of the timely destruction of the bridge at Dong Ha cannot be overstated - both in terms of Ripley's personal heroism, and the impact it had on the entire communist offensive. Those who ponder alternative history could easily argue that had the NVA been able to secure the bridge and the town at that time, the unfortunate end of the Republic of Vietnam on April 30th, 1975 might have been markedly speeded up.

Built by U.S. Navy Seabees in 1967, the bridge was a two-hundred-meter concrete and steel leviathan. Its destruction required deliberate planning, intellect and guts. Mostly guts. Ripley would provide all three, as he needed to distribute five hundred pounds of dynamite on the structure's underside.

Making a dozen-odd trips between the southern bank of the river and the belly of the bridge, each time he shuttled roughly forty pounds of explosives as he swung, hand-over-hand, out to the various spans and stringers, all the while exposed to enemy fire from the northern side. Placement of the dynamite and requisite wiring took more than two hours. With the rigging complete, and without fanfare, Smock and Ripley blew the bridge.

The details of this feat are riveting. The Third Battalion was composed of four rifle companies. Two of them, as well as Captain Ripley, spent the night before Easter Sunday at an abandoned combat base just west of Dong Ha. The NVA knew they were there, for they pounded the compound all night long with heavy artillery fire. The rounds came screaming in four or five a minute. The Vietnamese got little sleep. Ripley got none.

As the day dawned with an overcast sky, Ripley went out and examined the shell craters. The artillery fire was being directed away from the camp toward Dong Ha. He called his radio man to give a report to his own headquarters. Nha, a young baby-faced Vietnamese, approached with his long-range whip antenna waving back and forth. In the months they had fought together, the two had become inseparable. Neither knew the other's language well, but facial expressions and a common danger made words unnecessary. By that time Nha could read Ripley's mind.

Ripley grabbed the handset. Headquarters relayed the orders, "Fall back on Dong Ha and defend the bridge. I'll give you more information when I can." Binh's bodyguard, a powerfully built, rough individual who was known as "Three-fingered Jack," appeared and told Ripley that Binh wanted him at his command post. Jack was one of those quiet, alert veterans who command respect - he was a fearful enemy, and a welcome ally.

Binh had decided to deploy the two immediately available companies along the south bank of the Cua Viet River. One company would cover the main bridge used by the north-south traffic along Highway 1. It had been built by the Sea Bees five years earlier to carry the heaviest American weapons and equipment, including tanks. The other company would cover a much older bridge just upstream that could only carry light equipment. Binh told his Marines to dig their holes deep. There would be no fall back positions. They had to hold the riverbank.

The two companies formed a column, with Binh and Ripley leading the way, and headed for the bridge. Another radio message warned, "No time for questions, expect enemy tanks. Out." When they reached Highway 9, which ran along the south riverbank and intersected with Highway 1 at Dong Ha, it was clogged with thousands of refugees and, even worse, deserters by the hundreds. All of them had only one thought in mind - to get as far away as possible, as quickly as possible.

Binh's radio contact informed him that the rest of his battalion, plus a regular Army of the Republic of Vietnam (ARVN) tank battalion of about forty tanks, would rendezvous with them one mile west of the town. The medium tanks would be somewhat outgunned by the heavier Soviet T-54s, but they were certainly better than no tank support at all. The tank battalion commander, an ARVN lieutenant colonel, was waiting at the rendezvous point with his American adviser, Major James Smock. The former was less than enthusiastic about staying around and required constant urging to cooperate.

Nha approached Ripley. It was headquarters calling again. "Our outposts can hear the tanks coming. They are traveling in the scrub terrain just off the roadway, but sooner or later they are going to have to get back on Highway 1 to cross the bridge."

"Don't we have any aircraft up, to tell us how many?" Ripley asked.

"None yet. Low ceiling."

"Come on. We must have a thousand feet here."

"Believe me, pal, we are doing all we can. Every fire base up there is catching it and some have gone under. You have to hold the bridge and you have to do it alone. There is nothing here to back you up with."

Ripley's American adviser contact continued to give him bad news. Practically all resistance north of the bridge had been wiped out, which was probably the source of the ARVN deserters clogging the road along with the refugees. Then came the final blow. "We finally got a spotter plane in the air. They have tanks and armored personnel carriers stretched along Highway 1 for miles. Must be at least two hundred."

Ripley shouted back, "We can't stop that many. We have to blow the bridge at Dong Ha." At first his superior on the radio hesitated. The top brass back in Saigon wanted to save the bridge, but in the end Ripley's logic prevailed. A weary voice responded, "You are right. We can't authorize it, but you have to blow that bridge. Get moving that way and we will send some demo up to you."

As they approached Dong Ha, they saw the results of the destructive firepower of the enemy's heavy artillery. Corpses lay dismembered and forgotten along the roadside. Dead livestock and overturned carts were strewn in all directions. Then the artillery started again, with countless guns firing together, and shells exploding all over the town - but only the town. It was being blasted off the map. Everything came to a halt along the highway.

The tank column could not go forward and it could not stay where it was. They backed off to the west and swung around to the southeast and entered what was left of the town from the south. The shelling alternately intensified and then thinned out. At the outskirts, the tank commander refused to go any further, but after more arguments agreed to let two tanks accompany the dynamiters. As a parting remark, Binh told Ripley to send a message to his superiors. "There are Vietnamese Marines in Dong Ha. We will fight in Dong Ha. We will die in Dong Ha. As long as one Marine draws a breath of life, Dong Ha will belong to us." A hundred yards from the south end of the bridge Ripley, Smock and Nha prepared to go on alone.

Captain Ripley studied the bridge through his binoculars. It was built simply but massively. The bridge's basic strength lay in the steel I-beam girders that held up the superstructure. They ran longitudinally - that is, in the direction that the traffic would flow. Each girder stood three feet high, and the flanges extended three to four inches on either side of the vertical member. There were six of them across, with about three feet between them. With all that steel, Ripley thought to himself, the SeaBees could have built a battleship.

These hundred-foot long girders sat on top of massive, steel-reinforced concrete piers that rose twenty or thirty feet out of the river. At both sides of the river, the hundred-foot spans connected with the abutments. In thickness, the piers ran between five and six feet. They would easily have withstood any explosive power then available. The trick was to set the explosives in such a way as to knock one set of girders off the piers, thus dropping a hundred-foot span into the river - no small task, but possible if a soldier had the proper training. Fortunately, Captain Ripley had received the necessary training at Ranger School.

Ripley surveyed the scene directly in front of him. Along the near river bank, two companies of Binh's Marines were dug in. Across the river on the north side, there were thousands of NVA troops infesting the area. Halfway down his slope sat a bunker built up with sand bags left over from some previous battle.

The three stood up and made a dash for the bunker. As they ran, the fire from the north side increased in intensity and accuracy. They dove for the bunker just in time. Several shots thudded into the sand bags right in front of them. Ripley decided to leave Nha here, where he could make reports to headquarters just as easily without being exposed to any more danger than necessary.

He then attracted the attention of a squad leader at the river bank. Through sign language, he asked him to provide cover for the last leg of the journey to the bridge abutment. In a short period of time, Binh's Marines had a steady base of fire hitting NVA positions on the north bank.

The two officers broke from cover and ran straight for the bridge. Again the fire increased as they neared their objective. A heavy tank machine gun kicked up a spray of dirt in front of them. Ripley drove himself harder and harder. When he safely reached the bridge abutment, he almost collapsed from the exertion. He wondered how much longer he would have to keep going.

The explosives were waiting for them, about a dozen pine boxes and an equal number of canvas haversacks. Ripley read the stencil on the three-foot boxes: DEMOLITION-TNT. Each box contained 150 blocks of what looked like gray industrial soap. The haversacks contained plastic explosives to be used in conjunction with the TNT.

Ripley decided to cut the girders loose at the first pier, a hundred feet from the abutment. His problems began immediately. The SeaBees, to prevent sabotage to the under section of the bridge, had constructed a chain-link fence on the river side of the abutment topped with three coils of razor wire which Ripley would have to crawl over.

He chose to work on the downstream side of the bridge. Most of the infantrymen on both banks had dug in upstream, where they had more open space. The Marine captain climbed the fence and grabbed the bottom flanges of the I-beam. He then swung his feet up and hooked them on the flange.

He began to inch himself along the beam, and in doing so his legs took a beating. The razor wire sliced numerous cuts into his legs, which bled profusely. Through the wire he went. He was sweating heavily. The sweat rolled into his cuts and they began to burn. At last, he was through the wire.

With ninety feet to go, Ripley let his feet drop free and proceeded by hand-walking down the girder, swinging forward hand over hand. Arriving at the pier, he made an attempt to catapult himself up into the space between the outboard girder and the next one upstream. His legs would not cooperate. His energy was gone. Hanging only by his hands, they began to ache. Either he would flip up between the two beams soon, or he would fall into the river. Once again he tried, and almost made it. On the third try his heels caught the flanges. Then he twisted around until his body was spread-eagled between the two beams. He set the two haversacks of satchel charges, and crawled on his elbows and knees back to Major Smock and the fence.

The major passed the first two boxes of TNT and two more haversacks through the razor wire, which cut the major's hands and arms. Spread-eagled between the two girders, Ripley placed the boxes on the flanges and dragged the load, which weighed more than 180 pounds, back to the pier, where he set the charges to the first boxes of explosives.

Once more he dropped down, holding onto the bottom flanges with only his hands. He would swing back and forth, build momentum, leap, grab, catch the heels and then muscle into the channel opening between the next two girders. When his legs and lower body fell below the beams, the communist riflemen fired up into the steel girders, with rounds ricocheting all over. Nothing hit him. Once up into the channel he was safe.

For the next two hours, Ripley worked his way back and forth setting the charges. When he finished, he crawled back through the razor wire, dropped to the ground, and lay there for a while gasping for breath. Yet he had only accomplished the first part of the heroic undertaking. The exhausted Marine had to go out there again and set the detonators.

Ripley would have preferred to use electrical blasting caps and wire, but none were to be found - only the old-fashioned percussion caps and primer cord. To make things more difficult, they could not find any crimpers. Ripley had to crimp the caps onto the cord with his teeth. Since the shiny cylinders would explode if gripped too hard in the wrong place, a slight miscalculation would blow his skull apart. He remembered that back in Ranger School an instructor had placed a detonator inside a softball and set it off. The explosion had blown the cover, stuffing and string all over the place.

Carefully he placed the cap into his mouth, with the open end out, and put the primer cord in. He slowly bit down. It worked. The second time would be easier, but he had to fight off overconfidence - so he remembered the softball. Now the Marine captain was ready to go back out again.

This time the enemy was waiting for him. He crawled through the razor wire and dropped below the girder. The communists immediately opened fire, far heavier than before with hundreds of rounds bouncing off the girders. Over and over he prayed, "Our Lord Jesus Christ and His Blessed Mother, Jesus and Mary, get me there! Jesus and Mary, get me there..."

Just as he reached the upstream box of TNT, a tank shell hit the girder about two feet away. The angle was too flat and it bounced off and exploded on the south bank with a violent crash. The vibrations almost knocked him into the river. He set the detonator into the plastic explosive and lit the other end of the cord with a match. He had measured enough cord to allow about thirty minutes.

The girders of the Dong Ha Bridge were three feet high and about three feet apart. Ripley worked his way over to the downstream side and repeated the process and then hand-walked back to the fence. He realized that he had exceeded all normal human endurance, so again he turned to God and His Mother. "Jesus and Mary, get me there! Jesus and Mary, get me there..." He climbed back through the razor wire once more and fell to the ground near the abutment in a bloody heap. He was so tired that he could hardly lift his arm.

The major tapped him on the back. "Look what I found. But you won't need them now." He pointed to a box of electrical detonators. Ripley looked at the caps and realized that he had to go through the ordeal under the bridge once again. He had always been taught to rig up a backup charge if one was available. At this point, the substance of a man takes over. His moral integrity triumphs. In fact, throughout the entire ordeal, it was the guiding principle. So he returned again, simply because doing the job right demanded it.

While Ripley was again risking his life crawling around underneath the Dong Ha Bridge setting up the backup charges, Smock ran a couple of boxes of TNT down to the smaller bridge and ran back again. Ripley had completed the wiring and lay on the ground next to the abutment, too tired to move. Painfully he pulled himself up and, with a roll of detonating wire hung over his shoulder, staggered along with Smock back to the bunker where Nha was waiting. The South Vietnamese Marines unleashed a barrage of fire to cover them, yelling encouragement as they went, "Dau-uy Dien! Dau-uy Dien!" (Captain Crazy! Captain Crazy!)

At the bunker Ripley was glad to be reunited with Nha. He looked around for a way to trigger the explosion, since they had no blasting box. Nearby was a burned-out truck, but the battery appeared to be in good condition. Ripley tried several combinations to set off the explosives. Nothing worked. The terrible thought of failure came over him.

The captain would have to warn headquarters in order to give time for others to regroup farther south. He would stay with the Third Marine Battalion. Binh would never pull back. He had already made that clear. The battle-scarred warrior would die at his post with no forethought of death. From across the river, Ripley heard the tanks starting up. The massive assault was ready to begin.

Then the bridge blew. The shock waves came before the noise. The noise arrived, growing louder and louder in a series of explosions that became one huge roar. The entire hundred-foot span dropped into the river, leaving a huge gap in the bridge. The time fuses had done their job after all.

These were not the deeds of a regular man. His bravery was not some gut reaction or a counterpunch to a blow struck by an enemy. His actions in that three-hour window - with the world collapsing around him - were deliberate, willful, and premeditated. Every ounce of his spiritual and physical fiber was focused on mission accomplishment. Anything less, and he surely would have failed. Exhausted prior to the start, when he was finished he was way past empty - but by sheer will, he had prevailed.

STREET WITHOUT JOY

Hué City

"Hue City was the site of one of the most glorious chapters in Marine Corps history in which the Marines killed 5,113 enemy troops while suffering 147 dead and 857 wounded... but the Marines never got proper credit for Hue, for it was ultimately overshadowed by My Lai."
- Robert D. Kaplan

As the former imperial capital, Hué was for many Vietnamese the cultural center of their country. Both the South Vietnamese Army and Viet Cong troops for the most part refrained from any show of force in the immediate vicinity, or in the city itself. With a sort of unspoken truce in effect, Hué afforded both sides a certain respite from the war. With a wartime population of about 140,000 persons,

Hué retained much of its prewar ambience. Divided by the Huong or "Perfume" River, the city continued to emit a sense of both its colonial and imperial pasts. It was, in effect, two cities.

North of the river, the three-square-mile Citadel with its ramparts and high towers gave the appearance of a medieval walled town. Built by the Emperor Gia Linh in the early nineteenth century, it contained the former imperial palace with its large gilt and dragon-decorated throne room. Within the Citadel walls lay formal gardens and parks, private residences, market places, pagodas, and moats filled with lotus flowers. Buddhist bells and gongs as well as the chant of prayers resounded through its streets.

South of the river lay the modern city. Delineated by the Perfume River and the Phu Cam Canal into a rough triangle, southern Hué was about half the size of the Citadel. The university, the stadium, most government administrative buildings, the hospital, the provincial prison, and various radio stations were all in the new city. Attractive Vietnamese schoolgirls dressed in the traditional Ao Dai bicycled or walked along stately Le Loi Boulevard, paralleling the riverfront. The Cercle-Sportif, with its veranda overlooking the Perfume River, evoked memories of the former French colonial administration.

In January of 1968 as the Tet season approached, a certain uneasiness lay over the city. The cancellation of the Tet truce and the enemy attacks on Da Nang and elsewhere in southern I Corps dampened the usual festive mood of the holiday season. On January 30th, Brigadier General Ngo Quang Truong, the commanding general of the 1st ARVN Division, canceled all leaves and ordered his units put on full alert. Most of the troops, however, were already on leave and unable to rejoin their units. Moreover, the only South

Vietnamese forces in the city itself were the division staff, the division Headquarters Company, the Reconnaissance Company, a few support units, and Truong's personal guard - the elite "Black Panther" Company. The division headquarters was in the walled Mang Ca military compound, which was self-contained in the northeast corner of the Citadel. General Truong positioned the Black Panthers on the Tay Loc airfield in the Citadel, about a mile southwest of the division compound. In the southern city, the U.S. maintained a MACV compound in a former hotel which served as a billet and headquarters for the U.S. advisory staff to the 1st ARVN Division.

Although allied intelligence reported elements of two NVA regiments, the 4th and 6th, in Thua Thien Province, there was little evidence of enemy activity in the Hué sector. Indeed, the 1st ARVN Division dismissed any conjecture that the enemy had either the "intent" or "capability" to launch a division-size attack against the city. U.S. order of battle records listed the 6th NVA headquarters and its 804th Battalion in the jungle-canopied Base Area 114, about twenty to twenty-five kilometers west of Hué. One battalion, the 806th, was supposed to be in the "Street Without Joy" area in Phong Dien District, thirty-five kilometers northeast of Hué. American intelligence officers believed the remaining battalion, the 802nd, to be about twenty kilometers south of the city or with the regimental headquarters in Base Area 114. According to the best allied information, the 4th NVA Regiment was in the Phu Loc area near Route 1 between Phu Bai and Da Nang.

Unknown to the allies, both enemy regiments were on the move towards Hué. The 6th NVA had as its three primary objectives the Mang Ca headquarters compound, the Tay Loc airfield, and the imperial palace, all in the Citadel. South of the Perfume River, the 4th NVA was to attack the modern city. Among its objective areas were the provincial capital building, the prison, and the MACV advisors compound. The two regiments had nearly two hundred specific targets in addition to the primary sites, including the radio station, police stations, houses of government officials, the recruiting office, and even the National Imperial Museum. The target list contained detailed intelligence to the extent of naming suspected government sympathizers and their usual meeting places.

On January 30th, some of the enemy shock troops and sappers entered the city disguised as simple peasants. With their uniforms and weapons hidden in baggage, boxes, and under their street clothes, the Viet Cong and NVA mingled with the Tet holiday crowds. Many donned ARVN uniforms, and then took up predesignated positions that night to await the attack signal. By this time the 6th NVA Regiment was only a few kilometers from the western edge of the city.

At this point the 6th NVA divided into three columns, each with its particular objective in the Citadel. At 2200, about four kilometers southwest of Hué, the commander of the 1st ARVN Division Reconnaissance Company, First Lieutenant Nguyen Thi Tan, was on a river surveillance mission with about thirty men, when a Regional Force company to his east reported that it was under attack. Remaining under cover, Lieutenant Tan and his men observed the equivalent of two enemy battalions filter past their positions, headed toward Hué. Tan radioed this information back to the 1st Division. The two battalions were probably the 800th and 802d Battalions of the 6th NVA.

Despite Tan's warning, the enemy troops continued toward Hué unmolested. In the enemy command post to the west of the city, the NVA commander waited for word that the attack had begun. At approximately 0230 on 31 January a forward observer reported, "I am awake, I am looking down at Hué... the lights of the city are still on, the sky is quiet, and nothing is happening." Anxiously, the NVA officers looked at one another, but no one voiced their doubts. A few minutes later, the observer came back on the radio and announced that "the assault was underway."

At 0233 a signal flare lit up the night sky above Hué. At the Western Gate of the Citadel a four-man North Vietnamese sapper team, dressed in South Vietnamese Army uniforms, killed the guards and opened the gate. Upon their flashlight signals, lead elements of the 6th NVA entered the old city. In similar scenes throughout the Citadel, the North Vietnamese regulars poured into the old imperial capital.

The first U.S. Marines to bolster the South Vietnamese in the city were soon on their way. They were from the 1st Battalion, 1st Marines, part of Task Force X-Ray, the new command just established at the Marine base at Phu Bai, about eight miles south of Hué. As part of Operation Checkers, the Task Force X-Ray commander, Brigadier General Foster "Frosty" C. LaHue had opened his command post on 13 January. Two days later, as planned, he took over responsibility for the Phu Bai base from the 3rd Marine Division. LaHue, who had been at Da Nang until that time serving as the 1st Marine Division assistant division commander, had barely enough time to become acquainted with his new TAOR, let alone the fast-developing Hué situation. This was true as well for most of his commanders and units at Phu Bai.

With several changes making the original Checkers plan unrecognizable by the eve of Tet, LaHue had under him two regimental headquarters and three battalions. These were the

5th Marines under Colonel Robert D. Bohn with its 1st and 2nd Battalions, and the 1st Marines under Colonel Stanley S. Hughes, with its 1st Battalion in the Phu Bai sector. While Colonel Bohn had arrived with Task Force X-Ray on the 13th, Colonel Hughes did not reach Phu Bai until 28 January. The 1st Battalion, 1st Marines, under Lieutenant Colonel Marcus J. Gravel, began making its move from Quang Tri at about the same time. His companies C and D had reached Phu Bai on the 26th while his Company B and Headquarters Company came three days later. The battalion's remaining company, Company A, deployed on the 30th.

On 30 January the 1st Marines assumed from the 5th Marines responsibility for the Phu Bai area of operations as far south as the Truoi River. At the same time, Colonel Hughes took formal operational control of his 1st Battalion. Companies B, C, and D of the 1st Battalion, 1st Marines had already relieved the 2nd Battalion, 5th Marines at various bridges along Route 1 and other key positions in this northern sector. When Company A arrived on the 30th, it became the Phu Bai reserve or "Bald Eagle Reaction Force."

In the meantime, 2/5 had moved into the Phu Loc sector and took over that area south of the Truoi River and as far east as the Cao Dai Peninsula. 1/5 remained responsible for the rest of the Phu Loc region, extending to the Hai Van Pass.

In the Phu Loc area on 30 January, at about 1730, a Marine reconnaissance patrol codenamed "Pearl Chest" inserted about three thousand meters south of the town of Phu Loc and observed a North Vietnamese company moving north armed with three .50-caliber machine guns, AK-47s, and two 122mm rockets. Pearl Chest set up an ambush, killing fifteen of the enemy troops. The North Vietnamese fell back and surrounded the Recon Marines, who called for assistance. Both air and the artillery battery attached to 1/5 at Phu Loc responded to the request. The fixed-wing aircraft, however, could not get a fix on the enemy troops and were unable to assist.

At that point, about 1930, Lieutenant Colonel Robert P. Whalen, the 1st Battalion commander, sent his Company B to relieve the Recon team. As the relieving company approached the ambush site they heard Vietnamese voices, and someone threw a grenade at them. In return, the Marines hurled grenades of their own and then moved toward where they had heard the commotion. The enemy was no longer there, and the Marine company advanced cautiously. Lieutenant Colonel Whalen then asked Colonel Bohn, the 5th Marines commander, for reinforcements so as not to uncover his defenses at Phu Loc itself.

At the direction of Colonel Bohn, Lieutenant Colonel Ernest C. Cheatham, Jr., the 2/5 commander, sent his Company F to reinforce the 1st Battalion. Captain Michael P. Downs, the Company F CO, later recalled that the North Vietnamese ambushed his company as it moved into the 1st Battalion sector. At approximately 2300 on the 30th, about one thousand meters southeast of the Cao Dai Peninsula along Route 1, enemy troops opened up on the Marine company from the railroad tracks which paralleled the road with both automatic and semi-automatic weapons, killing one Marine and wounding three. After the initial burst the NVA broke contact, and the Marine company secured a landing zone to evacuate the wounded. Company F then returned to the 2nd Battalion perimeter.

By midnight on the 30th the engagement south of Phu Loc was about over. The Marine command did not want to commit any more troops, and ordered the Recon Team to break out and move to the north. Lieutenant Colonel Whalen then directed his Company B to return to Phu Loc, which it did without incident. The results of this activity were one Marine dead and five wounded, and sixteen enemy dead - with fifteen killed initially by the Recon Team, and another by Company B. Colonel Bohn, the 5th Marines commander, believed that this action prevented a full-fledged attack upon Phu Loc itself.

On the night of 30-31 January, at the same time the North Vietnamese struck Hué, the Marines had their hands full throughout the Phu Bai area of operations. Enemy rockets and mortars struck the Phu Bai airstrip and Communist infantry units hit Marine Combined Action units throughout the region. At the key Truoi River Bridge, an NVA company attacked the South Vietnamese bridge security detachment and the nearby Combined Action Platoon. Lieutenant Colonel Cheatham ordered Captain G. Ronald Christmas, the Company H commander, to relieve the embattled CAP unit. The Marines caught the enemy force beginning to withdraw from the CAP enclave and took it under fire. Seeing an opportunity to trap the North Vietnamese, Cheatham reinforced Company H with his Command Group and Company F, which by this time had returned from its abortive venture to Phu Loc.

With his other companies in blocking positions, Cheatham hoped to catch the enemy against the Truoi River. While inflicting casualties, the events in Hué were to interfere with his plans. Company G departed for Phu Bai as the Task Force reserve, and later that afternoon the battalion lost operational control of Company F. Years later Captain Downs remembered the company "disengaged... where we had them (the NVA) pinned up against a river, moved to the river and trucked into Phu Bai." With the departure of

Company F, the NVA successfully disengaged and Companies H and E took up night defensive positions. According to the casualty box score, the Marines of 2/5 in this engagement killed eighteen enemy troops, took one prisoner, and recovered sundry equipment and weapons including six AK-47s, at a cost of three Marines killed and thirteen wounded.

While the fighting continued in the Truoi River and the Phu Loc sectors, 1/1 had begun to move into Hué city. In the early morning hours of 31 January, after the rocket bombardment of the airfield and the initial attack on the Truoi River Bridge, Task Force X-Ray received reports of enemy strikes all along Route 1 between the Hai Van Pass and Hué. All told, the enemy hit some eighteen targets from bridges, Combined Action units, and company defensive positions. With Company A, 1st Battalion, 1st Marines as the Phu Bai reserve, Colonel Hughes directed Lieutenant Colonel Gravel to stage the company for any contingency. At 0630, Colonel Hughes ordered the company to reinforce the Truoi River Bridge.

The truck convoy carrying the company was escorted by two Army "Dusters," trucks armed with four .50-caliber machine guns, one at the head and the other at the rear of the column. When the convoy reached its destination, there were no ARVN troops to meet them. On their way south on Route 1, the company had passed several Combined Action units, whose troops told them "boo-coo" VC were moving towards Hué, but none had been hit, and all bridges were up. Batcheller then received orders from Lieutenant Colonel Gravel to reverse his direction, either to reinforce an Army unit north of Hué, or to go to the assistance of a Combined Action unit just south of Phu Bai. In any event, this new mission was short-lived. About one-half hour later the company again received another set of orders, presumably from Task Force X-Ray, "to proceed to the Hué Ramp area... to investigate reports that Hué City was under attack."

As the Marine company approached the southern suburbs of the city, they began to come under increased sniper fire. In one village, the troops dismounted and cleared the houses on either side of the main street before proceeding. The convoy then crossed the An Cuu Bridge, which spanned the Phu Cam canal, into the city. Caught in a murderous crossfire from enemy automatic weapons and B-40 rockets, the Marines once more clambered off the trucks and tanks. Sergeant Alfredo Gonzalez, a twenty-one-year-old Texan and acting 3rd Platoon commander, took cover with his troops in a nearby building. When enemy machine gun fire wounded one Marine in the legs, Gonzalez ran into the open road, slung the injured man over his shoulder, and despite being hit himself by fragments of a B-40 rocket, returned to the relative safety of the building. Responding to orders from Captain Batcheller, Gonzalez rallied his men, and the column was again on the move.

This time the Marine convoy only advanced about two hundred meters before Communist snipers again forced them to stop. The enemy was on both sides of the road with a machine gun bunker on the west side of the road. A B-40 rocket killed the tank commander in the lead tank. At that point Sergeant Gonzales, on the east side of the road with some men of his platoon, crawled to a dike directly across from the machine gun bunker. With his Marines laying down a base of fire, Gonzales jumped up and threw four grenades into the bunker, killing all the occupants.

As the Marine company cautiously made its way northward in the built-up area, Captain Batcheller maintained sporadic radio contact with Lieutenant Colonel

Gravel at Phu Bai. For the most part, however, he heard on his artillery and air radio nets nothing but Vietnamese. The convoy reached a causeway or elevated highway in the middle of a large cultivated area, and once again came under enemy sniper fire. Batcheller went to the assistance of a fallen man and was himself wounded seriously in both legs. Gunnery Sergeant J. L. Canley, a giant of a man at six feet, four inches tall and weighing more than 240 pounds, then took command of the company.

As Company A engaged the enemy on the outskirts of Hué, Colonel Hughes, the 1st Marines commander, requested permission from General LaHue to reinforce the embattled company. The only available reinforcements were the command group of the 1/1 and Company G of 2/5, which earlier that morning had become the Phu Bai reaction force in place of Company A. Lieutenant Colonel Gravel remembered that there was no intelligence on the situation in Hué, and that his own battalion was strung out in the Phu Bai sector with elements still at Quang Tri. He had never met Captain Charles L. Meadows, the Company G commander, until "that first day." Gravel said the only planning he was able to accomplish was to give the order, "Get on the trucks, men."

Crossing the An Cuu Bridge, Lieutenant Colonel Gravel's relief column reached Company A in the early afternoon. With the linking up of the two forces, Gravel kept the tanks with him but sent the trucks and the wounded, including Captain Batcheller, back to Phu Bai.

With the tanks in the lead, and Company A, the battalion headquarters group, and Company G following in trace, Gravel's makeshift command made its way toward the MACV compound, arriving there about 1515. By this time, the enemy attackers had pulled back their forces from the immediate vicinity of the compound.

Leaving Company A behind to secure the MACV compound, the Marine battalion commander took Company G, reinforced by the three tanks from the 3rd Tank Battalion and a few South Vietnamese tanks from the ARVN 7th Armored Squadron, and attempted to cross the main bridge over the Perfume River. Gravel left the armor behind on the southern bank to provide direct fire support. The American M48s were too heavy for the bridge, and the South Vietnamese tankers in light M24 tanks refused to go.

As the Marine infantry started across, an enemy machine gun on the other end of the bridge opened up, killing and wounding several Marines. One Marine, Lance Corporal Lester A. Tully, who was later awarded the Silver Star for his action, ran forward, threw a grenade, and silenced the gun. Two platoons successfully made their way to the other side. They turned left, and immediately came under automatic weapons and recoilless rifle fire from the Citadel wall. The enemy was well dug-in and firing from virtually every building in Hué city north of the river. Among the casualties on the bridge was Major Walter D. Murphy, the 1st Battalion S-3 or operations officer, who later died of his wounds. Captain Meadows remembered that he lost nearly a third of his company, either wounded or killed, "going across that one bridge and then getting back across that bridge."

By 2000, 1/1 had established defensive positions near the MACV compound and a helicopter landing zone in a field just west of the Navy LCU Ramp in southern Hué. On that first day, the two Marine companies in Hué had sustained casualties of ten Marines killed and fifty-six wounded. During the night, the battalion called in a helicopter into the landing zone to take out the worst of the wounded.

At 0700, Gravel launched a two-company assault supported by tanks towards the jail and provincial building. As a M79 grenadier from Company G, 5th Marines recalled: "We didn't get a block away (from the MACV compound) before we started getting sniper fire. We got a tank... got a block, turned right and received 57mm recoilless which put out our tank." The attack was stopped cold, and the battalion returned to the MACV compound.

In the meantime Marine helicopters had completed a lift of Company F, 2/5 into southern Hué. Although coming under machine gun fire from the Citadel walls across the river shortly after 1500, the Marine CH-46s carrying the company landed south of the LCU Ramp with minimum difficulty.

In southern Hué, the Marines made some minor headway and brought in further reinforcements. The 1st Battalion finally relieved the MACV radio facility that morning and later, after a three-hour fire fight, reached the Hué University campus. Although the NVA had dropped the railroad bridge across the Perfume River during the night, they left untouched the bridge across the Phu Cam Canal. About 1100, H 2/5 crossed the An Cuu Bridge over the canal in a "Rough Rider" armed convoy.

Enemy snipers opened up on the Marine reinforcements as the convoy, accompanied by Army trucks equipped with quad .50-caliber machine guns and two Ontos, entered the city. Near the MACV compound, the Marines came under heavy enemy machine gun and rocket fire. The Army gunners with their 'quad .50s' and the Marine Ontos, each with six 106mm recoilless rifles, quickly responded.

As the 1st Battalion began to clear its objective area, Lieutenant Colonel Gravel had only one infantry company, Company A, now under First Lieutenant Ray L. Smith, who had relieved the wounded Captain Batcheller. On the morning of the 4th its first objective was the Joan of Arc School and Church, only about one hundred yards away. Smith's Marines found themselves engaged in not only building-to-building, but room-to-room combat against a determined enemy.

In the school building, Sergeant Alfredo Gonzalez' 3rd Platoon secured one wing, but came under enemy rocket fire from across the courtyard. The Marine sergeant dashed to the window and fired about ten LAAWs to silence the enemy, but then a B-40 rocket shattered the grilled pane and struck him in the stomach, killing him instantly. Lieutenant Smith credited Gonzalez for taking out two enemy rocket positions before he was killed, and he was later awarded the Medal of Honor for his actions here as well as those on 31 January.

Once the school was secure, Smith's Company A maneuvered to the sanctuary which lay among a grove of trees and houses. Gravel wistfully recalled that it was "a beautiful, beautiful, church." As the troops advanced upon the building the NVA threw down grenades, killing or wounding several Marines. According to the battalion commander, "They (the enemy soldiers) were up in the eaves, and we couldn't get them out." Reluctantly, Gravel gave the order to fire upon the church. Marine mortars and 106mm recoilless rifles pounded the building. In the ruins, the battalion found two European priests, one Belgian and one French, both unhurt, but according to Gravel, "absolutely livid," that the Marines had bombarded the building. Believing he had little choice in his decision, Gravel thought the clerics in their dark clothing were fortunate to escape with their lives as the troops were braced to shoot at anyone in a black uniform.

The Marines in Hué began to adapt to the street fighting, so different from the paddies and jungle of the Vietnamese countryside in their previous sectors. As Captain Christmas of the 2nd Battalion later observed, "street fighting is the dirtiest type of fighting I know." Although one Marine fire team leader agreed with Christmas that "it's tougher in the streets," he also remarked, "it beats fighting in the mud... you don't get tired as quickly when you are running and you can see more of the damage you're doing to the enemy because they don't drag off their dead."

With little room to outflank the enemy, the battalion had to take each building and block "one at a time." According to Cheatham, "we had to pick a point and attempt to break that one strong point... and then we'd work from there." After a time Cheatham and his officers noted that the enemy defended on every other street. The Marines would take one street, and usually push through the next row of houses fairly quickly and then hit another defensive position.

Supported by the four tanks from the provisional platoon of the 3rd Tank Battalion which arrived with 1/1 on the 31st and a platoon of Ontos from the Anti-Tank Company of 1st Tank Battalion, the Marine infantry advanced methodically against stubborn enemy resistance. Cheatham had reservations about the employment of the tanks in his sector. He later commented, "you couldn't put a section of tanks down one of those streets. The moment a rank stuck its nose around the corner of a building, it looked like the Fourth of July." The enemy opened up with all the weapons in its arsenal from B-40 anti-tank rockets to machine guns. One tank sustained over 120 hits and another went through five or six crews. The battalion commander observed that when the "tankers came out of those tanks, they looked like they were punch drunk."

The Marine infantry commanders were much more enthusiastic about the Ontos with its six 106mm recoilless

.

rifles. Despite its thin skin, the vehicle was a big a help as any item of gear not organic to the battalions. Colonel Hughes later commented, "If any single supporting arm is to be considered more effective than all others, it must be the 106mm recoilless rifle, especially the M50 Ontos..." Hughes believed that the mobility of the Ontos made up for the lack of heavy armor protection, and that its plating provided the crew with sufficient protection against enemy small arms fire and grenades. From ranges of three to five hundred meters, the 106mm recoilless rifles rounds routinely opened four square meter holes or completely knocked out an exterior wall. Even at distances of one thousand meters, the recoilless rifles proved effective.

Although both battalions encountered moderate to heavy enemy resistance on the 5th, Lieutenant Colonel Cheatham's 2nd Battalion, 5th Marines made somewhat faster progress. Company G secured the main hospital building after a ninety-minute firefight supported by an M48 tank, 106mm recoilless rifles, and 3.5-inch rockets. The Marines removed the civilian patients as best they could from the line of fire, killed four NVA soldiers, and took thirty wounded prisoners. For the day, the three companies of the battalion accounted for over seventy North Vietnamese dead and forty captured enemy weapons.

Two Marine tanks came up to support the attack. One of the tanks took two direct hits from B-40 rockets but continued to fire. In addition, the Marines expended over one hundred 81mm mortar shells, sixty recoilless rifle rounds, and four E8 CS launchers in support of the assault on the headquarters. Wearing their gas masks, the tired Marines of Company H, in mid-afternoon, finally overwhelmed the NVA defenders in the provincial headquarters. They killed twenty-seven enemy soldiers, took three prisoners, and captured an assortment of enemy small arms and ammunition. The company had sustained one dead and fourteen wounded.

The province headquarters had served as a symbol for both the NVA and the Marines in the modern city. A now-frayed flag of the Viet Cong National Liberation Front had flown from the flagpole in the courtyard of the provincial building since the initial NVA takeover of the city. Immediately after the capture of the headquarters, two Marines rushed into the courtyard and hauled down the enemy ensign. Gunnery Sergeant Frank A. Thomas vaulted through a hole in the wall and ran to the flagpole clutching an American flag. As a CBS television crew filmed the event, Thomas raised the Stars and Stripes on the pole. According to Thomas, "We never knew exactly where the flag came from, but when we said we wanted an American flag to raise, one of our Marines produced one a very few minutes later." For this one time, the Marines ignored the MACV directive that forbade the display of the U.S. flag without the South Vietnamese national banner beside it.

On the morning of 7 February, both Marine battalions renewed their offensive. On the right flank, Cheatham's battalion with two companies on line and one in reserve made rapid progress. According to the battalion's entry for the day in its after-action report, "it became quite obvious the enemy had retreated leaving bodies and weapons behind." On the left flank 1/1 also moved forward, but at a slower pace, and met pockets of heavy resistance. The NVA knocked out an Ontos supporting the battalion with a B-40 rocket, killing the driver and wounding the vehicle's commander. After a firefight, a platoon from Company B retrieved the damaged vehicle, evacuated the wounded Marine, and recovered the body of the dead man.

By 10 February, despite some desperate efforts by isolated groups of NVA and the occasional sniper, the two Marine battalions had reached their objectives. With the Marines in control south of the Perfume River and the NVA still holding fast in the Citadel north of the river, Hué was now indeed two cities. Three days earlier, North Vietnamese sappers had blown the main bridge across the Perfume, literally dividing the city in two. Marine engineers destroyed the Le Loi Bridge at the end of Le Loi Street to prevent the enemy from bringing reinforcements into southern Hué from the west. At the same time, Marines from 1/1, reinforced by Company G, had secured the northern end of the wrecked An Cuu Bridge over the Phu Cam Canal. Lieutenant Colonel Cheatham and the remaining companies of the 2nd Battalion prepared to cross the Phu Cam and enter a new area of operations south of the city.

In clearing the modern city, the Marines took a heavy toll of the enemy, but at a high cost to themselves. The Americans had accounted for over one thousand enemy dead, took six prisoners, and detained eighty-nine suspects. Marine casualties included thirty-eight dead and about 320 wounded. Company H had been particularly hard hit. Every officer, including Captain Christmas, and most of the staff NCOs had sustained wounds. Corporals were now squad leaders. One Marine from Company G observed, "We would start getting new guys and it just seemed that every time we got new guys we would lose them just as fast." Another Marine from the same unit remarked, "The stink - you had to load up so many wounded, the blood would dry on your hands. In two or three days you would smell like death itself."

Although the battle for southern Hué was largely over, the fight for the Citadel had just begun. While the Marines cleared the new city, the South Vietnamese offensive in the Citadel had faltered. In the first days of the campaign the 1st Battalion, 3rd ARVN Regiment had cleaned out much of the northwest corner of the old city while the 1st ARVN Airborne Task Force, just south of the 1st Battalion, attacked from the Tay Loc airfield towards the western wall. To the east, the 4th Battalion, 2nd ARVN Regiment advanced south from the compound they occupied toward the former imperial palace grounds, which was enclosed within its own walls and moats. The battalion made excellent progress until enemy resistance stiffened about halfway toward the objective. By 4 February, the 1st ARVN Division reported that it had killed nearly seven hundred NVA troops in the Citadel.

On the night of 6-7 February, the NVA counterattacked. Using grappling hooks, fresh North Vietnamese troops scaled the southwestern wall and forced the 2nd Battalion, 4th ARVN to fall back with heavy losses to the Tay Loc airfield. That afternoon, the cloud cover lifted enough for South Vietnamese Air Force fixed-wing aircraft to drop twenty-five 500-pound bombs on the now NVA-occupied southwest wall of the Citadel.

With the clearing of southern Hué by the 1st Marines, General Cushman prepared to bring more forces into the fight for the entire city. The 1st Battalion, 5th Marines were about to expand Marine operations in Hué City into the old Citadel to reinforce the ARVN. Simultaneously, the Marine command attempted to improve the coordination for artillery, naval gunfire, and other supporting arms for the Citadel fighting. Earlier, the 1st Field Artillery Group (FAG) at Phu Bai, the artillery command for Task Force X-Ray, deployed four 155mm howitzers of Battery "W," 1st Battalion, 11th Marines to firing positions at Gia Le, about three thousand meters west of Phu Bai, to improve supporting fires for the forces in Hué.

On 10 February the 1st FAG commander, Lieutenant Colonel John P. Barr, ordered two officers on his staff to the Citadel area as forward observers. One of the officers, First Lieutenant Alexander W. Wells, Jr., the S-2 (intelligence officer) on the FAG staff, remembered that he received word that morning that the colonel wanted to talk to him. Barr informed Wells that he had volunteered the young lieutenant for a twenty-four-hour mopping-up mission in the Citadel.

Shortly after 1630 on the 10th, Wells and his radio operator flew by helicopter to the Tay Loc airfield in the Citadel where the Marine lieutenant was to provide support to the 2nd Battalion, 4th ARVN and the Black Panther Company, which had just retaken the field. As the aircraft approached Tay Loc, the enemy took it under sniper fire. The two Marines leaped out of the hovering craft and ran into a Quonset hut, near the airfield tower, and "full of Australians (advisors to the Vietnamese) playing cards and drinking scotch."

Upon Wells reaching the division headquarters, General Truong briefed him upon his new assignment as a forward observer supporting remnants of an ARVN Airborne battalion pinned down in a forward area. Wells remembered that he was shocked to learn that 1/5 had not arrived yet and that he and his radioman would be the only Americans in actual combat with the ARVN. The Vietnamese general pointed out to Wells, on a large wall map, the location of his designated outpost - which was surrounded by enemy troops. Truong explained the Vietnamese unit required "his 'big guns' immediately to break the siege." According to Wells, 'Truong emphasized that the Emperor's Palace of Perfect Peace and the Royal City itself were in a strict no-fire zone.

After the briefing two ARVN Rangers escorted the Marine lieutenant and his radioman through the dark streets and alleyways to the ruins of a Buddhist pagoda about five hundred meters west of the Dong Ba tower. It took him about three hours to negotiate the half-mile distance from the Mang Ca compound to the pagoda. Inside and around the courtyard of the temple and only a short distance from the Imperial Palace were about one hundred Vietnamese troops who were surrounded by North Vietnamese forces. Given his ominous circumstances, Lieutenant Wells nicknamed his refuge the "Alamo," and for the next two weeks called in Marine supporting artillery and naval gunfire from ships off the coast.

The support elements of the Vietnamese Marine Task Force reached Phu Bai on the night of 10 February from Saigon and began preparations to move the 1st Battalion into the Citadel. On the morning of 11 February, U.S. helicopters started the helilift of the Vietnamese Task Force headquarters and 1st Battalion into the Citadel. Low ceiling and drizzle forced a halt in the air movement of the Vietnamese Marines with only the task force headquarters and one company of the 1st Battalion in the old city. General LaHue proposed that the remainder of the battalion be trucked to southern Hué and then board LCM (landing craft mechanized) for the trip downriver to a landing site north of the Citadel. The Marines would then move on foot into the city. The Vietnamese commander refused, as he did not feel that either route was sufficiently secured. It would be two days before additional units of the Vietnamese Marine task force joined the one company in the Citadel.

In the meantime 1/5 began to go into the old city. Shortly after 1045 on 11 February, Marine CH-46 "Sea Knight" helicopters lifted three platoons of Company B from the Phu

Bai airfield to the Mang Ca compound in the Citadel. Enemy gunfire wounded the pilot of the helicopter carrying the 3rd Platoon, forcing him to abort the mission and return to Phu Bai with the troops still on board. Later that day Company A, with five tanks attached from the 1st Tank Battalion, embarked in a Navy LCU at the ramp in southern Hué. After their relatively uneventful cross-river passage, the Marine company and tanks joined the two platoons of Company B at the 1st ARVN Division headquarters.

That same day Major Robert H. Thompson, the commanding officer of 1/5, and his command group accompanied his remaining companies from the Phu Loc sector to Phu Bai. Only ten days before Colonel Bohn, the regimental commander, had chosen Thompson - who had served with him before as a battalion operations officer - to take over the battalion after the wounding of its previous commanding officer. Before assuming command of the battalion Thompson, a lieutenant colonel selectee, had been the III MAF embarkation officer. The NVA had prepared a rather undignified assumption of command ceremony for the new battalion commander. Thompson recalled, "the moment I stepped off the helicopter at Phu Loc we received mortar incoming. My first fifteen minutes with 1/5 was spent at the bottom of a muddy fighting hole with my baggage and several Marines piled on top of me."

On the morning of 13 February, 1/5 moved out of the Mang Ca compound with two companies abreast - Company A on the left and Company C on the right, with Company B in reserve. From the outset the Marines encountered enemy elements of squad and platoon size in well prepared positions and bunkers dug in built up areas and along the Citadel walls. In Major Thompson's words, "within fifteen minutes... all Hell broke loose. There was no Airborne unit in the area and Company A was up to their armpits in NVA." Under fire from automatic weapons, fragmentation grenades, B-40 rockets, mortars, and AK-47s, Company A sustained thirty-five casualties within minutes. Among the wounded was the company commander.

At that point Major Thompson ordered his reserve, Captain Jennings' Company B, to relieve Company A. First Lieutenant Scott A. Nelson's Company C resumed the attack with Company B on its left flank. With two tanks in the lead, Company C advanced about three hundred meters before heavy enemy fire from an archway tower along the Citadel's eastern wall leading to the Dong Ba Bridge once more stopped the Marines. The NVA had dug in at the base of the wall there and tunneled back underneath this structure. While protected by the thick masonry from allied supporting fires, the enemy could use the archway to bring further reinforcements into the Citadel. With the Marine battalion about seventy-five meters short of its original proposed line of departure, Colonel Hughes radioed Major Thompson to hold his positions.

The next morning the battalion resumed the attack. Offshore, Navy cruisers and destroyers opened up with their 5-inch and 8-inch guns. Marine 8-inch and 155mm howitzers from firing positions at Phu Bai and Gia Le added to the bombardment. For the first time in several days, the cloud cover lifted for a brief period and Marine F-4B Phantoms and F-8 Crusader jets flew support missions.

The Marine attack stalled. On the right, Company C advanced about one hundred yards, destroyed an NVA rocket position, and captured an enemy soldier who walked into the company lines. But on the left flank, Company B made no progress against the enemy-occupied tower. After several futile attempts to take the tower, Major Thompson ordered both companies back into night defensive positions.

Earlier that day, Captain Myron "Mike" Harrington's Company D had reverted to Thompson's command. Harrington brought two of his three platoons to the LCU ramp in southern Hué for transportation down river to the Citadel. At the ramp, there were two LCUs, but fully loaded with supplies for the 1st Battalion. Harrington squeezed on board one of the craft with his headquarters group and one infantry squad. Although taking fire from NVA gunners on the Citadel wall, the Navy craft safely made the trip across the river. Harrington and his small force jumped off and waited for the LCUs to make a return trip with the rest of the company.

At the LCU ramp, the remaining two platoons boarded the Navy craft to join their company commander and his small detachment. Again as the LCUs made their way across the Perfume, NVA gunners took them under fire. On the opposite shore, two Marine 4.2-inch mortars responded with both high explosive and CS shells. A sudden shift of wind brought the gas fumes back on the Navy boats, blinding and choking both the sailors and Marines. The two LCUs then returned to the southern ramp, and the ship commanders decided against another attempt to cross the river. Fortunately after several hours a Navy Swift Boat arrived with three Vietnamese junks in row. Armed with a mounted .50-caliber machine gun, the Swift Boat commander agreed to take the Marines on board the junks and tow the small convoy to the other side. After the Swift Boat left the junks at a point offshore, the Marines rowed them to the northern landing site where an impatient Captain Harrington was waiting for them.

On the 15th, Marine artillery and naval gunfire once more hit the enemy positions. Under the pounding this time, part of the tower gave way. With another break in the cloud cover, two Marine A-4 jets darted in under the gray skies and dropped 250 and 500-pound bombs on the target. Backed both by tanks and Ontos, the Company D Marines pressed forward with Company C protecting their right flank. The North Vietnamese defended their positions tenaciously, and Major Thompson ordered Company B, which had been in reserve, once again into the attack. After six hours of hard fighting, including hand-to-hand combat, Harrington's 1st Platoon established a foothold at the base of the tower. According to one account, Marine Private First Class John E. Holiday made a "one-man charge" against an enemy machine gun bunker on the wall, firing his machine gun from the hip, "John Wayne style." The rest of the company followed him and captured the tower.

The capture of the tower came at no small cost. Thompson's battalion lost six men killed and sustained more than fifty wounded, while claiming twenty enemy dead. That night, Captain Harrington left one squad in the tower and established his CP in a damaged house below the wall. Then, in a surprise night attack, the NVA retook the tower for a brief period. According to Harrington, the Marine squad fell back without orders and the company commander at the base of the tower suddenly saw North Vietnamese soldiers crawling over the rubble of the tower. Laying down a base of fire from his defensive positions, Captain Harrington led another squad in a counterattack. The tower finally remained in Marine hands.

For the next few days the 1st Battalion met the same close-quarter resistance from the enemy. In contrast to the enemy in southern Hué, the battalion discovered that the

NVA units in the Citadel employed better city-fighting tactics, improved the already formidable defenses, dug trenches, built roadblocks and conducted counterattacks to regain redoubts which were important to their defensive scheme. Major Thompson later observed that the older city consisted of row after row of single-story, thick-walled masonry houses jammed close together and occasionally separated by alleyways or narrow streets, and the Marines encountered hundreds of naturally camouflaged, mutually supporting, fortified positions.

Thompson countered the enemy fixed defenses with heavy artillery, naval gunfire, liberal use of riot control agents, and when the weather permitted, fixed-wing support. The Marine battalion commander depended largely on his unit's own firepower, especially his mortars and automatic weapons, and the tanks and Ontos that reinforced his battalion. He placed both the tanks and Ontos under the control of the attached tank platoon commander. The infantry provided a screen while the mobile Ontos or tanks furnished direct fire support. In order to enhance observation the tank or Ontos commander, together with the infantry commander, would reconnoiter the target area - generally a building blocking the Marine advance. The tank or Ontos commander then returned to his vehicle and prepared to move forward at full speed as the infantry Marines laid down a heavy volume of fire. Upon reaching a position where fire could be placed on the target, the vehicle commander halted his vehicle and fired two or three rounds into the target and then reversed his direction and returned quickly to friendly lines.

At first the M48 tank's 90mm guns were relatively ineffective against the concrete and stone houses - shells occasionally even ricocheted back upon the Marines. The tank crews then began to use concrete-piercing fused shells which resulted in excellent penetration, and walls were breached with two to four rounds. Although casualties among the Ontos and tank crews were high, the tanks themselves withstood with relatively little damage direct hits by the enemy RPG rounds. Major Thompson compared the tankers to the "knights of old sallying forth daily from their castles to do battle with the forces of evil," and one Marine rifleman said, "If it had not been for the tanks, we could not have pushed through that section of the city. The NVA seemed to have bunkers everywhere."

In the Citadel, Major Thompson had decided on another tack to get his battalion moving again. On the afternoon of the 20th he held a conference with his company commanders and suggested the possibility of a night attack. According to Thompson, most of the company commanders were not very enthusiastic. "They were willing to try, but I could see that their hearts were not in it." He understood their reluctance - they had endured a great deal during the past two weeks. On the other hand a few days earlier he had given his reserve company, Company A, to First Lieutenant Patrick D. Polk. In a brief period Polk had revived the morale of the company, which had taken horrendous casualties on the first day of action in the Citadel. According to the plan, a platoon from Company A was to seize three key facilities, including the two-story administrative building, flanking the North Vietnamese positions during the night. At first light, the rest of the battalion was to launch the general attack.

The 2nd Platoon of Company A, led by Staff Sergeant James Munroe, moved out at 0300 from the company perimeter. Divided into three ten-man teams, the Marines captured all three buildings with only minimum resistance by the enemy. Major Thompson later speculated that the North Vietnamese withdrew from the buildings during the night to sleep elsewhere. In the morning about daybreak the enemy troops started to move back, providing a "turkey shoot" for the Marines of Company A. According to one of the Marine enlisted men, "Hell, the first thing in the morning we saw six NVA just standing on the wall. We dusted them all off." This threw the NVA into utter confusion and gave the other companies the spirit they needed to continue the attack with zest. Despite the initial success, the North Vietnamese defended the ground within the zone of action with tenacity.

The end, however, was in sight. On the 21st L 3/5 relieved Company B, which received a well-earned rest, and the following morning the 1st Battalion prepared for the final assault on the southern wall. Lieutenant Polk carefully briefed Company A, which this time was to be in the vanguard of the attack. At 0930 the Marines once more pushed forward, and except for some scattered snipers and an occasional mortar round the enemy seemingly had melted away. Upon reaching the southeastern wall of the Citadel Lance Corporal James Avella broke out a small American flag from his pack and fastened it to a sagging telegraph pole. The battalion's after-action report documented this event with the phrase, "an element of Company A hoisted our National Ensign."

Upon securing the wall, Major Thompson ordered the new company under his command, Company L, to capture the southern gate and the immediate area outside the Citadel leading to the bridge across the river. The assault was a classic combined arms effort that could not have been executed better on a blackboard. The sun was out for the first time in two weeks, Marine fixed-wing aircraft dropped napalm within two hundred meters of the advancing troops, and by 1800 the Marine battalion had succeeded in attaining all of its objectives.

For the 1/5 Marines in the Citadel, except for isolated skirmishes, their last significant action occurred on the 22nd with the seizure of the southeast wall and its approaches. Major Thompson had hoped to participate in the taking of the Imperial Palace, but as he later ruefully observed, "For political reasons, I was not allowed to do it. To save face, the Vietnamese were to retake the Forbidden City." Marine tanks, Ontos and recoilless rifles, however, provided direct support for the assault on the palace.

The battle had cost all sides dearly. Marine units of Task Force X-Ray sustained casualties of 142 dead and close to 1,100 wounded. U.S. advisors with the 1st ARVN Division in Hue reported 333 South Vietnamese Army troops killed, 1,773 wounded, and thirty missing in action. According to the U.S. Marine advisors with the Vietnamese Marine task force in Hué, the Vietnamese Marines suffered eighty-eight killed, 350 wounded, and one missing in action. The 1st Cavalry Division (Airmobile) listed casualties of sixty-eight killed and 453 wounded for their part in the battle, while the 1st Brigade, 101st Airborne showed six dead and fifty-six wounded in its battle account. Thus, all told, allied unit casualties totaled more than six hundred dead and nearly 3,800 wounded and missing. Obviously the enemy did not escape unscathed. Allied estimates of NVA and VC dead ranged from 2,500 to 5,000 troops. Captured Communist documents admitted to 1,042 killed, and an undisclosed number of wounded.

Given both the resources that the North Vietnamese put into the battle and the tenacity with which they fought, it was obvious that the Hué campaign was a major component of the entire Tet offensive. According to an enemy account, the North Vietnamese military command in planning the offensive took into consideration that the U.S. and South

Vietnamese had concentrated their forces in the north, expecting an attack along Route 9. It viewed Hué a weak link in the allied defenses in the northern two provinces.

Once in Hué, the North Vietnamese were there to stay. The Communists established their own civil government and their cadres rounded up known government officials, sympathizers, and foreigners including American civilians and military personnel in the parts of the city they controlled. After the recapture of Hué, South Vietnamese authorities exhumed some three thousand bodies thrown into hastily dug graves. In all probability, these were the victims of the Communist roundups. Although the North Vietnamese admitted tracking down and punishing "hoodlum ringleaders," they claimed most of the reported civilian deaths were the result of happenstance, exaggerations by the South Vietnamese, or were caused by the allies. The true sufferers in the battle were the people of Hué.

Some estimates held that over eighty percent of the structures in the city sustained damage or were destroyed. Out of a population of about 140,000, more than 116,000 people were homeless and 5,800 were either dead or missing - Hué was a devastated city.

One of the most valiant - and overlooked - chapters in Marine Corps history had come to a close.

ONE TOUGH MARINE

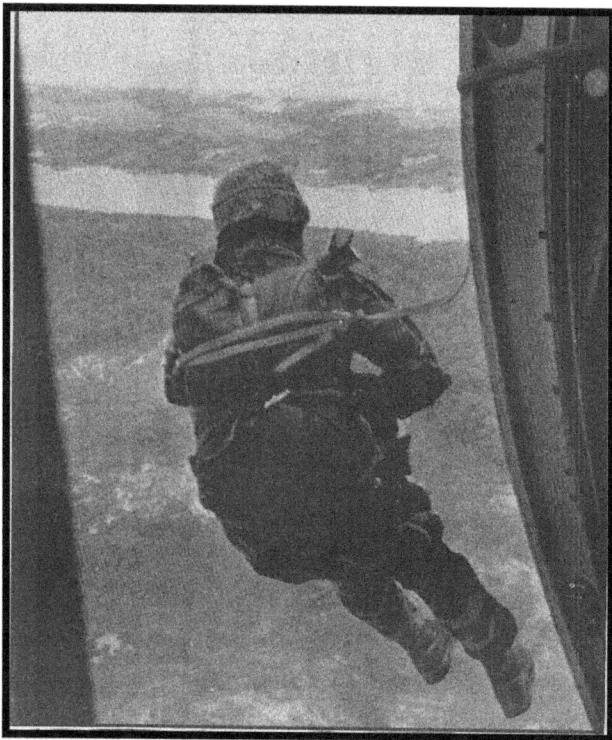

First Sergeant Donald N. Hamblen

"The mind is the limit. As long as the mind can envision the fact that you can do something, you can do it, as long as you really believe one hundred percent."
- Arnold Schwarzenegger

Throughout the history of the United States Marine Corps many men have stood proudly at attention before their fellow Marines as they were recognized for making some lasting contribution that bettered our Corps, or for having distinguished themselves in acts of great personal courage on the battlefield. Their individual achievements and moments of bravery have become a lasting part of our military history; these accomplishments serve as examples for all to follow. However, the pedestal of recognition on which these Marines have stood often seem to be made of a very delicate, shifting sand as their moments of recognition, being much too brief, are quickly eroded by time.

Occasionally, a very few Marines find themselves having to stand tall more than once, as again they are recognized for their sustained exemplary service or for their repeated demonstrations of personal courage during combat. These Marines are an extremely rare breed and they become our mentors and our heroes. Don "Woody" Hamblen is one such individual.

In 1950, an eighteen-year-old boy from the small New England farming community of Winthrop, Maine questioned whether or not he was man enough to "wear the Greens" and enlist in the ranks of the United States Marine Corps. He had heard that the Marines were tough and wanted to join "with the best." He went to boot camp at Parris Island, and got his first taste of combat serving in Korea from 1951 to 1952 as a rifleman assigned to D Company, 2nd Battalion, 5th Marine Regiment. While there, he was wounded twice in one day fighting against the North Korean Army.

Following the Korean War Hamblen chose service in the Marine Corps as his life's profession, and he learned his trade with the infantry at Camp Lejeune, at the Marine Barracks in New London Connecticut, and with the 1st Marine Brigade in Hawaii. Subsequently he returned to Camp Pendleton, California where he volunteered for duty with a Marine Force Reconnaissance Company and honed his skills at Army Airborne School, Navy Underwater

Swimmer's School, and Marine Corps Mountain Leadership School.

In September of 1962, while assigned as the pathfinder platoon sergeant for 1st Force Reconnaissance Company, Hamblen was making his two hundred and fifteenth parachute jump when strong, gusting winds caused him to drift down and away from his designated drop zone. His nonsteerable parachute canopy became entangled in a series of three sixty-nine-thousand-volt electrical power lines, and while helplessly suspended beneath those high-tension cables he swayed back and forth until his left jump boot made contact with a lower twelve-thousand-volt line, causing him to be electrocuted.

Hamblen's green parachute canopy sparked into a ball of orange and blue flames and melted as fire swept over him. His jump boot burned around his foot, and he fell more than thirty feet to the ground where he landed in a smoking heap. Badly burned, but still alive and fully conscious, he was taken by rescue helicopter to the base hospital where five days later his blackened left leg was amputated six inches below his knee.

But the story does not end here with this injured Marine staff sergeant being discharged from the Corps, receiving an artificial leg, and being left with the feeling of being someone less than whole. Instead this Marine non-commissioned officer embarked on a personal crusade to recover from what he believed was only a "temporary" handicap, and requested he be given the chance to prove he was physically fit and should be allowed to remain on active duty in the Corps. While he waited for an official reply he relearned how to stand, how to walk, and how to run. Hamblen soon demonstrated that he was able to scuba dive, and he also returned to parachuting. The story of his remarkable recovery soon became newsworthy, and he had the opportunity to represent our country's disabled veterans when he was invited to Washington, D.C., to attend the annual meeting of the President's Committee on Employment of the Handicapped. While there he met with the Commandant of the Marine Corps, General Wallace M. Greene, various senators and members of Congress, and was honored by President Lyndon Johnson. To these leaders and thousands of others, his remarkable recovery was an incredible example of personal courage.

After returning to full-duty status with 1st Force Reconnaissance Company in 1963, Hamblen became the first United States Marine wearing a prosthesis to be assigned to combat duty. He spent thirty consecutive months in Vietnam operating as a reconnaissance specialist and Studies and Observations Group (SOG) team advisor, operating in both the southern and northern regions of the country. And he was wounded again - twice.

His is the story of one dedicated individual who simple would not quit in the face of tremendous adversity, and who repeatedly demonstrated he had the courage, spirit, and self-determination to overcome his misfortune. Don Hamblen's extraordinary example of tenacity is personified in the phrase, "one tough Marine."

THE NEW BREED

Captain Brian Chontosh

"An imperfect plan executed immediately and violently is far better than a perfect plan next week."
– General George S. Patton

Meet Brian Chontosh. As a boy he attended Churchville-Chili Central School, Class of 1991. Proud graduate of the Rochester Institute of Technology. Husband and soon-to-be father. First Lieutenant in the United States Marine Corps. And a genuine hero. Chontosh was presented with the Navy

Cross, the second highest award for combat bravery the United States can bestow. That's a big deal.

But you didn't see it on the network news, and all you read in Brian's hometown newspaper was two paragraphs of nothing. Instead, it was more blather about some mentally defective MPs who acted like animals at Abu Ghraib Prison.

The odd fact about the American media during the Iraq War was it did not cover the American military. The most plugged-in nation in the world received virtually no true information about what its warriors were doing.

Oh, sure, there was a body count. We knew how many Americans fell. And we saw those same casket pictures day in and day out, and were almost on a first-name basis with the idiots who abused the Iraqi prisoners. And we knew all about improvised explosive devices and how we lost Fallujah and what Arab public-opinion polls said about us and how the world hated us.

We got a non-stop feed of gloom and doom. But we didn't hear about the heroes. The incredibly brave troops who honorably did their duty. The ones our grandparents would have carried on their shoulders down Fifth Avenue. The ones we completely ignored. Like Brian Chontosh.

During the march into Baghdad Brian Chontosh was a platoon leader rolling up Highway 1 in a Humvee when all hell broke loose. It was ambush city.

His young Marines were being cut to ribbons. Mortars, machine guns, rocket propelled grenades. And the kid out of Churchville was in charge. It was do or die, and it was up to him.

So he moved to the side of his column, looking for a way to lead his men to safety. As he tried to poke a hole through the Iraqi line his Humvee came under direct enemy machine gun fire. It was like shooting fish in a barrel, and the Marines were the fish.

And Brian Chontosh gave the order to attack. He told his driver to floor the Humvee directly at the machine gun emplacement that was firing at them. And he had the guy on top with the .50-cal unload on them.

Within moments there were Iraqis slumped across the machine gun and Chontosh was still advancing, ordering his driver now to take the Humvee directly into the Iraqi trench that was attacking his Marines. Over into the battlement the Humvee went and out the door Brian Chontosh bailed, carrying an M16 and a Beretta and 228 years of Marine Corps pride.

And he ran down the trench, with its mortars and riflemen, machineguns and grenadiers. And he killed them all.

He fought with the M16 until it was out of ammo. Then he fought with the Beretta until it was out of ammo. Then he picked up a dead man's AK47 and fought with that until it was out of ammo. Then he picked up another dead man's AK47 and fought with that until it too was out of ammo. At one point he even fired a discarded Iraqi RPG into an enemy cluster, sending attackers flying with its grenade explosion.

When he was done Brian Chontosh had cleared two hundred yards of entrenched Iraqis from his platoon's flank. He had killed more than twenty, and wounded at least as many more.

But that's probably not how he would tell it. He would probably merely say that his Marines were in trouble, and he got them out of trouble. Ooh-Rah, and drive on.

"By his outstanding display of decisive leadership, unlimited courage in the face of heavy enemy fire, and utmost devotion to duty, First Lieutenant Chontosh reflected great credit upon himself and upheld the highest

traditions of the Marine Corps and the United States Naval Service."

That's what the citation said. And that's what nobody heard about. That's what didn't make the evening news. Accounts of American valor were dismissed by the press as propaganda, yet accounts of American difficulties were heralded as objectivity. It makes you wonder if the role of the media is to inform, or to depress? To report, or to deride? To tell the truth, or to feed us lies?

But I guess it doesn't matter. We're going to turn out all right. As long as men like Brian Chontosh wear our uniform. He is proof that we still makes Marines like we used to.

CORPS CHRONOLOGY

1775 **10 Nov** - Continental Congress authorizes two battalions of American Marines; the birthday of the Corps.
28 Nov - Samuel Nicholas commissioned as Captain of Marines.

1776 **3 Mar** - Capt Nicholas and his Marines land at New Providence, Bahamas, seize military stores. And in January fight for Gen Washington at Princeton.

1778 **27 Jan** - Captain John Trevett leads twenty-six Marines in capture of Fort Nassau, Bahamas.
22 Apr - John Paul Jones and Lt Samuel Wallingford spike cannon at Whitehaven, England. Later, Wallingford is killed when Ranger defeats Drake.

1779 **28 Jul** - Capt John Welsh and thirteen Marines killed in assault on Fort George at Penobscot Bay, Maine.

1798 **11 Jul** - President John Adams signs Act establishing the U.S. Marine Corps.

1805 **27 Apr** - Lt Presley O'Bannon and seven Marines lead attack against Derna, Tripoli.

1814 **24 Aug** - In Battle of Bladensburg 114 Marines help defend Washington, D.C.

1815 **8 Jan** - Andrew Jackson, including Marines under Major Daniel Carmick, defeats British at New Orleans.
20 Feb - Capt Archibald Henderson leads Marines in Constitution's victory over Cyane and Levant.

1820 **8 Oct** - LtColComdt Anthony Gale cashiered.
17 Oct - Archibald Henderson appointed commandant, holds position for thirty-eight years until his death.

1824 **12 Mar** - Brevet Maj Robert D. Wainwright and Marines from Boston quell riot in state prison.

1832 **8 Feb** - Brevet Capt Alvin Edson leads attack at Quallah Battoo, Sumatra.

1836 **23 Jun** - ColComdt Henderson and 462 Marines report for duty in the Second Seminole War.

1837 **27 Jan** - Col Henderson wins Battle of Hatchee-Lustee River against Seminoles in Florida.

1846 **30 Jul** - 1stLt Jacob Zeilin leads Marine detachment ashore at Santa Barbara, California. Two weeks later Commo Robert Stockton, with 360 Marines and sailors, enters Los Angeles.

1847   **9 Mar** - Capt Alvin Edson leads Marine battalion ashore with Army forces at Veracruz, Mexico.

       13 Sep - In Mexico City, Marines help seize fortress of Chapultepec and next day occupy the National Palace on site of the Halls of Montezuma.

1853   **14 Jul** - Marines under Maj Jacob Zeilin land as Commodore Matthew Perry opens Japan.

1856   **20 Nov** - Marines in landing party that captures four barrier forts guarding way to Canton, China.

1859   **18 Oct** - 86 Marines under 1stLt Israel Green, USMC, and LtCol Robert E. Lee, USA, capture abolitionist John Brown at Harpers Ferry, Virginia.

1861   **21 Jul** - Battalion of 365 Marines led by Brevet Maj John G. Reynolds fights in Battle of Bull Run.

1862   **15 May** - Cpl John F. Mackie on board ironclad Galena is first Marine to receive the Medal of Honor.

1863   **26 Apr** - 250 Marines under Capt John L. Broome seize New Orleans custom house and city hall.

1865   **15 Jan** - 365 Marines in naval landing force attack Fort Fisher at Wilmington, North Carolina.

1871   **10 Jun** - Capt McLane Tilton leads 109 Marines in naval attack on Han River forts in Korea.

1880   **1 Oct** - John Philip Sousa appointed 17th leader of the Marine Band.

1885   **18 Jun** - Marines land in Panama to protect trans-isthmus railroad.

1898   **15 Feb** - 28 Marines among 250 Americans killed when cruiser Maine is blown up in Havana harbor.

       1 May - Admiral George Dewey destroys Spanish fleet in Manila Bay and Marines occupy Cavite Naval Station.

       10 Jun - 1st Marine Battalion led by LtCol Robert W. Huntington lands at Guantanamo Bay, Cuba. Sgt John Quick signals under Spanish fire to save Marine unit, receives Medal of Honor.

1899   **8 Oct** - Marines attack Filipino insurgents at Novaleta.

1900   **31 May** - Marines reach Chinese capital to defend Legation Quarter from Boxer rebellion.

       4 Aug - Marines in International Relief Force that marches out of Tientsin to lift siege of Peking. Pvt Dan Daly wins first of two Medals of Honor.

HARD CORPS ~ Legends of the Corps

1901	**28 Sep** - Maj "Tony" Waller takes out 314 Marines to destroy Filipino insurgents who slaughtered a U.S. Army company on Samar.
1903	**5 Nov** - Maj John Lejeune lands battalion to ensure Panama's independence from Colombia.
1906	**28 Sep** - Provisional Marine Brigade of 2,800 men lands at Havana; Marines stay until 1909.
1908	**12 Nov** - President Theodore Roosevelt removes Marines from warships, but six months later President Taft restores them.
1912	**22 May** - 1stLt Alfred Cunningham is first Marine aviator. **14 Aug** - Maj Smedley D. Butler leads Marines ashore, beginning intervention in Nicaragua. **4 Oct** - Marines fight at Coyotepe, Nicaragua.
1914	**21 Apr** - Marine regiments land at Veracruz, Mexico, to keep German guns from Mexican dictator.
1915	**28 Jul** - Marines land in Haiti beginning their longest Caribbean intervention. **18 Nov** - Maj Butler leads Marines in attack on Fort Riviere in Haiti, awarded his second Medal of Honor.
1916	**15 May** - Marine battalion begins occupation of Dominican Republic.
1917	**6 Jun** - 5th Marine Regiment sails for France.
1918	**6 Jun** - Marines advance into Belleau Wood against German machine guns. **29 Jun** - Marines from USS Brooklyn go ashore at Vladivostok, Siberia. **18 Jul** - Marines in vast Allied counter-offensive meet Germans south of Soissons. **3 Sep** - In Haiti, native leader Charlemagne Peralte starts revolt of "Cacos" against Marine rule. **12 Sep** - In France, 2d Division including Marine Brigade begins offensive in Saint Mihiel salient. **3 Oct** - 4th Marine Brigade assaults Blanc Mont in fierce fighting. **14 Oct** - Marine fliers 2dLt Ralph Talbot and GySgt Robert G. Robinson win Medals of Honor. **1 Nov** - Marine Brigade enters Meuse-Argonne. **10 Nov** - 5th Marines make night crossing of the Meuse River against German resistance.

1919	**31 Oct** - Marine Sgt Herman Hanneken and Cpl William Button sneak into "Cacos" camp and kill Peralte.
1924	**12 Jul** - Marine Brigade leaves Dominican Republic.
1927	**6 Jan** - Marines begin second Nicaraguan intervention, fight Augusto Sandino in the mountains.
	16 Mar - 4th Marines land at Shanghai to stay fourteen years.
1928	**6 Jan** - 1stLt Christian Schilt begins ten flights to aid besieged Marine patrol at Quilali, Nicaragua.
	8 Mar - Capt Merritt A. Edson sets out on epic Coco River patrol to hunt for Sandino.
1930	**25 Jul** - Lt Lewis Puller wins first of five Navy Crosses chasing Sandino guerrillas in Nicaragua.
1933	**2 Jan** - 5th Marine Regiment departs Nicaragua.
	14 Nov - Marines at Quantico, Virginia, begin work on "Tentative Landing Operations Manual."
	7 Dec - Navy Department creates Fleet Marine Force.
1934	**15 Aug** - Marines end intervention in Haiti.
1937	**7 Dec** - Marine Capt Evans Carlson goes to Yenan to observe Communist Chinese armies in action.
1941	**7 Jul** - 1st Provisional Marine Brigade lands at Iceland.
	27 Nov - 4th Marines leave Shanghai marking end of era.
	7 Dec - Marines try to fight back when Japanese attack U.S. Pacific fleet at Pearl Harbor. Marines at Tientsin and Peking are forced to surrender.
	10 Dec - Japanese defeat American garrison on Guam.
	23 Dec - Japanese overwhelm garrison on Wake Island.
	28 Dec - Most of 4th Marines move from Bataan to Corregidor Island in Manila Bay. Fortress island falls 6 May 1942.
1942	**9 Apr** - 105 Marines among Americans on Bataan Death March.
	1 Jun - First black Marines enlist in Corps
	7 Aug - 1st Marine Division lands on Guadalcanal.
	17 Aug - 2d Raider Battalion raids Makin Atoll.
	13 Sep - On Guadalcanal, Marines turn back Japanese attack in Battle of Edson's Ridge.
1943	**13 Feb** - Woman's Reserve program is announced; birthday of women Marines.
	21 Jun - 4th Raider Battalion lands on New Georgia.
	16 Sep - Maj Gregory "Pappy" Boyington shoots down five planes; he would claim 28, the most of any Marine.
	1 Nov - 3d Marine Division lands on Bougainville.

20 Nov - 2d Marine Division assaults Betio Island of Tarawa Atoll in Central Pacific.

26 Dec - 1st Marine Division lands on Cape Gloucester, New Britain.

1944 **1 Feb** - 4th Division's 23d and 24th Marines land on Roi and Namur of Kwajalein Atoll in Marshalls.

17 Feb - 22d Marines help Army seize Eniwetok Atoll.

15 Jun - 2d and 4th Marine divisions assault Saipan.

21 Jul - 3d Marine Division opens battle for Guam.

24 Jul - 2d and 4th Marine divisions land on Tinian, clear airfield from which "Enola Gay" will take off for Hiroshima a year later.

15 Sep - 1st Marine Division assaults Peleliu.

1945 **19 Feb** - 4th and 5th Marine divisions assault Iwo Jima, raise flag on Mount Suribachi four days later.

26 Mar - Iwo Jima secured. Marines suffer 25,851 casualties.

1 Apr - On Easter Sunday, U.S. Tenth Army, including 1st and 6th Marine divisions, lands on Okinawa.

18 Jun - LtGen Simon Buckner, USA, killed; Marine MajGen Roy Geiger takes command of Tenth Army.

30 Aug - 4th Marines land at Yokosuka on Tokyo Bay.

30 Sep - Marines of III Amphibious Corps start landing in North China, disarms 630,000 Japanese.

6 Oct - On Tientsin-Peiping road, Marines have first fire fight with Chinese Communists.

1946 **6 May** - Commandant Archibald Vandegrift tells Senate Naval Affairs Committee "the bended knee is not a tradition of our Corps."

1948 **23 May** - First Marines brought ashore by helicopter for amphibious exercise at New River, North Carolina.

18 Jul - Twelve Marines at consulate in Jerusalem begin modern Marine Security Guard program.

10 Nov - First eight enlisted women are sworn in as Regular Marines. The following summer, first black women Marines enlist.

1949 **18 Nov** - Corps orders all male Marines, regardless of race, be assigned to vacancies in any unit.

1950 **2 Aug** - 1st Marine Provisional Brigade lands at Pusan, South Korea.

15 Sep - 1st Marine Division makes assault landing at Inchon on west coast of Korea, retakes Seoul.

2 Nov - Marines engage Chinese Communists in North Korea near the Chosin Reservoir.

23 Nov - Thanksgiving Day, 7th Marines take Yudam-ni.

28 Nov - After repulsing eight Chinese divisions, Marines begin epic "breakout" on 1 December.

1951 **20 Jun** - 1st Marine Division reaches "the Punchbowl" in Korea.

1952 **28 Jun** - Congress sets Marine Corps' strength and gives commandant equal status on Joint Chiefs of Staff in matters of concern to the Corps.

1954 **10 Nov** - Marine Corps War Memorial dedicated next to Arlington National Cemetery.

1956 **7 Jan** - Marine Security Guard fights off mob at consulate in East Jerusalem.

8 Apr - Six recruits drowned in Ribbon Creek at Parris Island, South Carolina.

1958 **15 Jul** - 2d Marines land near Beirut and seize airport at Lebanese government's request.

1962 **20 Feb** - Marine LtCol John H. Glenn, Jr, orbits earth in first manned American space capsule.

15 Apr - Marine helicopter squadron (HMM-362) arrives in Mekong Delta of South Viet Nam.

1965 **8 Mar** - 9th Marine Expeditionary Brigade lands at Da Nang, South Viet Nam.

28 Apr - 6th Marines land in Dominican Republic.

27 Oct - Viet Cong raids wreck Marine aircraft at Marble Mountain and Chu Lai.

1967 **28 Feb** - PFC James Anderson Jr, is first black Marine to win the Medal of Honor.

1968 **20 Jan** - North Vietnamese open battle against 26th Marines for Khe Sanh.

31 Jan - Vietnamese Communists launch Tet offensive.

5 Jul - Americans give up Khe Sanh base.

1971 **25 Jun** - Last Marine ground troops leave Viet Nam.

1974 **19 Aug** - Marines defend embassy in Nicosia, Cyprus, after mob kills U.S. ambassador.

1975 **12 Apr** - Marines evacuate foreigners before Khmer Rouge seize Phnom Penh, Cambodia.

30 Apr - 4th Marines under Col Alfred M. Gray complete evacuation by helicopters from Saigon embassy and Tan Son Nhut airfield.

14 May - Marines board American container ship Mayaguez, which Cambodians had seized, and assault pirates on Koh Tang Island

10 Nov - Marine Corps celebrates its 200th birthday.

1979   **30 Oct** - Embassy Marines drive off crowd with tear gas in San Salvador, El Salvador.

4 Nov - Mob overruns embassy in Teheran, Iran; thirteen Marines are among sixty-five Americans taken hostage; fifty-two are held captive for 444 days.

21 Nov - In Islamabad, Pakistan, mob burns embassy as seven Marines defend building.

1980   **24 Apr** - Three Marines killed in desert accident during effort to rescue Teheran hostages.

12 May - Embassy Marines in San Salvador use tear gas to rescue U.S. ambassador from mob.

23 Jun - Marines land in Lebanon to evacuate civilians.

1983   **18 Apr** - One Marine among sixty-three killed when terrorists blow up U.S. Embassy in Beirut, Lebanon.

23 Oct - Terrorist truck-bomb blows up headquarters of 1st Battalion, 8th Marines at Beirut airport, killing 241 Americans, of whom 220 are Marines.

25 Oct - Marines and Army intervene in Grenada, West Indies.

1984   **31 Jul** - All Marines except embassy guard leave Lebanon after 533-day intervention.

1989   **20 Dec** - Marines are part of force that deposes Panama's dictator, Manuel Antonio Noriega, during Operation Just Cause

1990   **5 Aug** - Marines execute Operation Sharp Edge and land in Monrovia, Liberia, to evacuate civilians threatened by civil war.

25 Aug - 7th Marine Expeditionary Brigade begins Persian Gulf buildup for Operation Desert Shield.

1991   **4 Jan** - Marine helicopters evacuate 281 people from U.S. Embassy at Mogadishu, Somalia.

16 Jan - Operation Desert Storm begins as Marines fly in first waves of allied planes.

24 Feb - Marine 1st and 2d divisions, commanded by LtGen Walter Boomer, breach Iraqi line.

14 Apr - Marines ordered to Iraq-Turkey border to help multinational relief force protect Kurds.

30 Apr - Marines assist millions of homeless after cyclone kills 125,000 in Bangladesh.

1992   **9 Dec** - Marines land in Somalia to rescue foreigners.

1993   **20 Jun** - Marine unit returns to Mogadishu, Somalia, to maintain peace.

1994   **12 Apr** - Marines evacuate foreigners from Rwanda.

      20 Sep - Army and 1,900 Marines land in Haiti.

1995   **3 Mar** - Marines complete withdrawal of UN force from clan war in Somalia.

8 Jun - Team of forty Marines rescues Air Force pilot shot down over Bosnia-Herzegovina on 2 June.

1997   **30 May** - Marines help evacuate 2,500 from Kinshasa, Zaire.

2001   **6 Oct** – Marines deploy to Afghanistan for Operation Enduring Freedom

2003   **April 11** – I MEF conducts Operation Iraqi Freedom

2004   **Nov 8** – Marines assault insurgents in Fallujah, Iraq

RECOMMENDED READING

One Tough Marine
By Donald Hamblen & Bruce Norton
Line of Departure: Tarawa
By Martin Russ
Battle for Hue: Tet 1968
By Keith William Nolan
The Marine Raiders
By Edwin P. Hoyt
Wake Island
By Duane Schultz
Marine Sniper
By Charles Henderson
Guadalcanal Diary
By Richard Tregaskis
Marine!
By Burke Davis
A Marine Named Mitch
By Mitchell Paige
Marine Rifleman
By Wesley Fox
Small Unit Action in Vietnam: Summer 1966
By Francis J. West, Jr.
The Bridge at Dong Ha
By John Grider Miller
Victory at High Tide
By Robert Debs Heinl, Jr.
One Bugle, No Drums
By William B. Hopkins
The U.S. Marine Corps Story
By J. Robert Moskin

HARD CORPS ~ Legends of the Corps

ABOUT THE AUTHOR

Andy Bufalo retired from the Marine Corps as a Master Sergeant in January of 2000 after more than twenty-five years service. A communicator by trade, he spent most of his career in Reconnaissance and Force Reconnaissance units but also spent time with Amtracs, Combat Engineers, a reserve infantry battalion, and commanded MSG Detachments in the Congo and Australia.

He shares the view of Major Gene Duncan, who once wrote "I'd rather be a Marine private than a civilian executive." Since he is neither, he has taken to writing about the Corps he loves. He currently resides in Tampa, Florida.

Semper Fi!

316